FUNCTIONAL POLYMER BLENDS AND NANOCOMPOSITES

A Practical Engineering Approach

T0181072

FUNCTIONAL POLYMER BLENDS AND NANOCOMPOSITES

A Practical Engineering Approach

Edited by

**Gennady E. Zaikov, DSc, Liliya I. Bazylak, PhD, and
A. K. Haghi, PhD**

Apple Academic Press

TORONTO   NEW JERSEY

Apple Academic Press Inc. | Apple Academic Press Inc.
3333 Mistwell Crescent | 9 Spinnaker Way
Oakville, ON L6L 0A2 | Waretown, NJ 08758
Canada | USA

©2014 by Apple Academic Press, Inc.

First issued in paperback 2021

Exclusive worldwide distribution by CRC Press, a member of Taylor & Francis Group
No claim to original U.S. Government works

ISBN 13: 978-1-77463-324-3 (pbk)
ISBN 13: 978-1-926895-89-5 (hbk)

Library of Congress Control Number: 2014935853

Library and Archives Canada Cataloguing in Publication

Functional polymer blends and nanocomposites: a practical engineering approach/edited by Gennady E. Zaikov, DSc, Liliya I. Bazylak, PhD, A.K. Haghi, PhD.

Includes bibliographical references and index.
ISBN 978-1-926895-89-5 (bound)
1. Polymers. 2. Nanocomposites (Materials). I. Haghi, A. K., author, editor of compilation II. Zaikov, G. E. (Gennadiĭ Efremovich), 1935- author, editor of compilation III. Bazylak, Liliya I.,author, editor of compilation

QD381.F85 2014   547'.7   C2014-901859-2

Apple Academic Press also publishes its books in a variety of electronic formats. Some content that appears in print may not be available in electronic format. For information about Apple Academic Press products, visit our website at **www.appleacademicpress.com** and the CRC Press website at **www.crcpress.com**

ABOUT THE EDITORS

Gennady E. Zaikov, DSc
Gennady E. Zaikov, DSc, is Head of the Polymer Division at the N. M. Emanuel Institute of Biochemical Physics, Russian Academy of Sciences, Moscow, Russia, and Professor at Moscow State Academy of Fine Chemical Technology, Russia, as well as Professor at Kazan National Research Technological University, Kazan, Russia. He is also a prolific author, researcher, and lecturer. He has received several awards for his work, including the Russian Federation Scholarship for Outstanding Scientists. He has been a member of many professional organizations and is on the editorial boards of many international science journals.

Liliya I. Bazylak, PhD
Liliya Bazylak, PhD, is Senior Staff Scientist in the Physical and Chemistry of Combustible Minerals Department at the Institute of Physical and Organic Chemistry and Coal Chemistry at the National Academy of Sciences of Ukraine in Lviv. She is the author of more than 200 publications. Her scientific interests include physical chemistry, nanochemistry and nanotechnologies, and chemistry of high molecular compounds.

A. K. Haghi, PhD
A. K. Haghi, PhD, holds a BSc in urban and environmental engineering from University of North Carolina (USA); a MSc in mechanical engineering from North Carolina A&T State University (USA); a DEA in applied mechanics, acoustics and materials from Université de Technologie de Compiègne (France); and a PhD in engineering sciences from Université de Franche-Comté (France). He is the author and editor of 65 books as well as 1000 published papers in various journals and conference proceedings. Dr. Haghi has received several grants, consulted for a number of major corporations, and is a frequent speaker to national and international audiences. Since 1983, he served as a professor at several universities. He

is currently Editor-in-Chief of the *International Journal of Chemoinformatics and Chemical Engineering* and *Polymers Research Journal* and is on the editorial boards of many international journals. He is a member of the Canadian Research and Development Center of Sciences and Cultures (CRDCSC), Montreal, Quebec, Canada.

CONTENTS

LIST OF CONTRIBUTORS

O. I. Aksimentyeva
Department of Physical and Colloidal Chemistry, Ivan Franko National University of L'viv; D.Sc. in Chemical Sciences, the major staff scientist

L. I. Bazylyak
Physical-Chemistry of Combustible Minerals Department, Institute of Physical-Organic Chemistry and Carbon Chemistry named after L. M. Lytvynenko, National Academy of Sciences of Ukraine; Oxidizing Processes Chemistry Department; Dr. in Chemical Sciences, senior staff scientist

V. P. Dyakonov
Institute of Physics, Polish Academy of Science; Dr. in Chemical Sciences

Yu. M. Hrynda
Physical-Chemistry of Combustible Minerals Department, Institute of Physical-Organic Chemistry and Carbon Chemistry named after L. M. Lytvynenko, National Academy of Sciences of Ukraine; Oxidizing Processes Chemistry Department; post-graduate training student

O. Yu. Khavunko
Physical-Chemistry of Combustible Minerals Department, Institute of Physical-Organic Chemistry and Carbon Chemistry named after L. M. Lytvynenko, National Academy of Sciences of Ukraine; Oxidizing Processes Chemistry Department; Dr. in Chemical Sciences, junior researcher

G. I. Khovanets
Physical-Chemistry of Combustible Minerals Department, Institute of Physical-Organic Chemistry and Carbon Chemistry named after L. M. Lytvynenko, National Academy of Sciences of Ukraine; Oxidizing Processes Chemistry Department; Dr. in Chemical Sciences, junior researcher

V. I. Kopylets'
Physical-Mechanical Institute named after G. V. Karpenko, National Academy of Sciences of Ukraine; Dr. in Technical Sciences, senior staff scientist

S. A. Korniy
Physical-Mechanical Institute named after G. V. Karpenko, National Academy of Sciences of Ukraine; Dr. in Technical Sciences, senior staff scientist

A. S. Kun'ko
Department of Physical and Colloidal Chemistry, Ivan Franko National University of L'viv; Dr. in Chemical Sciences

A. R. Kytsya
Physical-Chemistry of Combustible Minerals Department, Institute of Physical-Organic Chemistry and Carbon Chemistry named after L. M. Lytvynenko, National Academy of Sciences of Ukraine; Oxidizing Processes Chemistry Department; Dr. in Chemical Sciences, senior staff scientist

Yu. G. Medvedevskykh
Physical-Chemistry of Combustible Minerals Department, Institute of Physical-Organic Chemistry and Carbon Chemistry named after L. M. Lytvynenko, National Academy of Sciences of Ukraine; Chief of the Oxidizing Processes Chemistry Department; D.Sc. in Chemical Sciences, the major staff scientist

V. I. Pokhmurskiy

Physical-Mechanical Institute named after G. V. Karpenko, National Academy of Sciences of Ukraine; D.Sc. in Technical Sciences, professor, corresponding member of National Academy of Sciences of Ukraine

O. V. Reshetnyak

Department of Physical and Colloidal Chemistry, Ivan Franko National University of L'viv; Department of Chemistry at Army Academy named after hetman Petro Sahaydachnyi; D.Sc. in Chemical Sciences, professor, head of the Department of Physical and Colloidal Chemistry

M. M. Yatsyshyn

Department of Physical and Colloidal Chemistry, Ivan Franko National University of L'viv; Dr. in Chemical Sciences, associate professor

G. E. Zaikov

Kinetics of Chemical and Biological Processes Division Institute of Biochemical Physics named after N. N. Emanuel, Russian Academy of Sciences D.Sc. in Chemical Sciences, head of the Department

I. M. Zin'

Physical-Mechanical Institute named after G. V. Karpenko, National Academy of Sciences of Ukraine; D.Sc. in Technical Sciences, the major staff scientist

PREFACE

The main attention in this collection of scientific papers is focused on recent theoretical and practical advances in polymers and nanocomposites. It consists of two parts: the first is devoted to fundamental theoretical investigations concerning the conformational and deformational demonstrations of polymers into solutions and melts, and the second is dedicated to some of the newest practical achievements in the fields of polymers, nanocomposites, and nanoparticles.

Conformation is the statistical property of macromolecules consisting of a number of N structural units, links, and position of which in a space one relative to the other is not inflexibly fixed by chemical bonds and assumes the possible random configurations. The number of possible configurations at $N > 1$ is so great that it permits the use of statistical methods at their analysis. That is why the conformation is a result of statistical averaging on all possible configurations of the macromolecule. For a long time it was noted and confirmed by computer modeling that the conformation of polymeric chain should to be described by self-avoiding random walks statistics $(SARW)$ but not Gaussian random walks statistics assuming the phantom behavior of polymeric chains. Nevertheless, Gaussian statistics is dominating at the analysis of thermodynamic, dynamic, and kinetic staining's of the macromolecules conformation. Undoubtedly, the works of Kuhn, Flory, de Jennes, Kirkwood, de Clause, and other researchers are and remain as fundamental and basic for understanding the theoretical sense and logical development of physical chemistry of polymers. However, the starting point of these works in a view of the Gaussian statistics, even in context of new a scaling idea, does not give the adequate quantitative description of the conformation properties of macromolecules that do not permit in full measure to estimate their role in different equilibrium and nonequilibrium processes. Development of the self-avoiding random walks statistics $(SARW)$ and its application to the analysis of thermodynamic, dynamic and kinetic staining's of the confirmative properties of macromolecules in solutions and melts is the "know-how" presented

Chapters 1–6. According to the SARW statistics presented in these chapters, the macromolecules' conformation possesses all thermodynamic properties: entropy, free energy, volume, pressure, etc. At the same time it is important to note that the contribution of the conformation in general entropy of random walks is negative. In this sense the macromolecules conformation is a form of its self-organization. As equilibrium property, the conformation is resistant to transition from more probable state into the less probable one. Hence, there are elastic properties of conformation, the measure of which is not only the pressure, but also the deformation's modules, namely Young's modulus and sifting's module. Introduction of free conformation energy into the structure of chemical potential of macromolecules gives a general approach to the analysis of the conformation role into equilibrium processes. Dynamic properties of the macromolecules described in these chapters are determined first of all by the reptation mechanism of their translational and rotation movement of the base of which is segmental movement, which is ordering to SARW statistics. Taking of the above-said into account permits obtaining the expressions for the diffusion coefficient of macromolecules into solutions and melts, which are quantitatively agreed with the experimental data. Additionally taking into account the elastic properties of the conformation in a form of equilibrium shifting module and peculiarities of kinetics of shear deformation gave the possibility to obtain the expressions for viscosity of polymeric solutions and melts and to describe its gradient dependence.

Chapters 7–12 are dedicated to some of the newest practical achievements in the field of polymers and nanocomposites and are covered by the different branches of knowledge: from the obtaining of nanoparticles with the use of new synthetic techniques to application of the obtained nanomaterials and nanocomposites in different fields of industry. For example, composites of conducting polymer and inorganic oxides are very attractive and are prospective materials for different branches of science, namely, for chemistry, physics, electronics, photonics, etc., due to synergetic effect which arises under the integration of the properties of oxide and conducting polymer. At present the titanium (IV) oxide (rutile and anatase crystalline modifications) is one among widely used inorganic oxides, which is applied for the chemical synthesis of such nanocomposites. The combination of the properties of nano-TiO_2 and polyaniline helps to solve problems successfully in the chemistry, physics, and electronics. Specific electronic structures of the nano-TiO_2 (as the n-type semiconductor) and polyaniline (as the electron's conductor in majority of the cases and as a p-type semiconductor under certain conditions) give the possibility to design the systems for different applications. Therefore, the main aim of the investigations presented in Chapter 7 was the synthesis and study of physico-chemical properties of such composite materials.

One way to extend the functionality of the organic materials with electronic conductivity is a chemical or electrochemical copolymerization of different classes of compounds, such as naphthalene derivatives and aromatic and heterocyclic monomers. In synthesis the nanocomposites based on the conduct of polymer and organic supramolecules, such as heteropolyacids or aminonaphthalene sulphonic acids acting both as molecular dopants and surfactants, attracts a rising interest because of the possibility to the nanotubes or nanofibers and polymer nanodispersions during polymerization. However, the results of the connection between structure and properties of copolymers of aniline with other type of aminonaphthalene sulphonic acids were sufficiently received. Chapter 9 deal with the studies of the amino naphthalene sulfonic acid nature effect on the structure and physical properties of their copolymers with aniline; namely, the authors studied the physical properties and structure of the copolymers based on polyaniline *(PANI)* and amino naphthalene sulfonic acids (ANSA) with different mutual position of substituents in naphthalene ring, namely, 1-amino-naphthalene-4-sulfonic acid (1,4-ANSA), 1-amino-naphthalene-8-sulfonic acid (1,8-ANSA), 1-amino-naphthalene-5-sulfonic acid (1,5-ANSA) and 1-amino-2-naphthol-4-sulfonic acid. It is shown that structure of isomeric amino naphthalene sulfonic acids has a significant effect on the thermal stability, magnetic susceptibility, and conductivity of the PANI-ANSA systems. The field dependence of magnetization confirms that obtained copolymers are the typical paramagnetic materials.

Chapter 8 is devoted to the simulation of corrosive dissolution of *Pt* binary nano-cluster in acid environment, of polymer electrolyte membrane fuel cells. It is well known that under the present catalytic electrode production for low temperature fuel cell, it is necessary to reduce their costs by the proposal of binary platinum nanoclusters Pt_nX_m (where X are the transition metals *Cr, Fe, Co, Ni, Ru*), while such nanoparticles may possess high catalytic properties. However, in practice any nanocatalyst effective work period undergoes the sufficient reduction as a result of its degradation during the corrosive dissolution in the acid environment. Thus, despite the enhanced catalytic properties of the binary nanoclusters, their stability and the corrosive resistance obey to be reduced for the prolonged work. The stored experimental results need an interpretation on atomic and molecular level in frames of different models, as a sequence of peculiarities of reaction ways on nanoparticles and under the influence of size effects in nanosystems. Meanwhile, the known theoretical approaches in such simulations lead to indefinite information, while the methods used on the basis of these approaches are unable for adequate description of the nanoparticle properties, because the lack of quantum behavior understanding during the interactions in such nanosystems. In Chapter 8, a model is proposed describing the corrosive dissolution of binary nanocluster surfaces for Pt_nX_m with core shell structure in acid environment of low temperature fuel cells in the presence of molecules and ions H_2O, Cl^-, OH^-, H_3O^+, which is based on calculations of the adsorption properties during the interaction of environment components with the surface as well as the activation barriers of platinum atoms dissolution by means of DFT methods. The model includes the surface structure, sizes, and forms of nanoclusters, including their chemical composition, which is able sufficiently to change catalytic activity, stability, and corrosive resistance of the nanocatalysts.

An exponential growth in the field of the fundamental and applied sciences connected with a synthesis of the nanoparticles of noble metals, and studies of their properties and practical application is observed for the last decades. Such rapid development of the scientific investigations in the above-said fields is caused first of all by the development of the instrumental and synthetic methods of obtaining and investigations of such materials in connection with their wide use in microelectronics, optics, catalysis, medicine, sensory analysis, etc. Silver nanoparticles are characterized by unique combination of the important physical-chemical properties, namely by excellent optical characteristics, by the ability to amplify the signal in spectroscopy of the combination dispersion, and also by high antibacterial properties. Among all the metals possessing the characteristic phenomenon of the surface plasmon resonance exactly, silver is characterized by the greatest efficiency of the plasmon's excitation that leads to the abnormally high value of the extinction coefficient of silver nanoparticles. Under conditions of modern tendency to the miniaturization and the necessity to improve the technological processes of the obtaining of new materials based on silver nanoparticles, there is a problem of their identification, which involves the cost of equipment and causes a search for alternative ways to find their average size and size distribution determination by other methods, in particular, by calculated ones with the use of the empirical equations and dependencies which are based on the property of the adsorption of the electromagnetic irradiation in *UV*/visible diapason by the sols of silver nanoparticles. In Chapter 10, on a basis of the comparative analysis of the references, the correlated dependencies between the optical characteristics of aqueous sols of spherical nanoparticles and their diameter have been discovered. As a result, the empirical dependencies between the values of the square wave frequency in the adsorption maximum of the surface plasmon resonance and average diameter of the nanoparticles were determined as well as between the values of the adsorption bandwidth on half of its height and silver nanoparticles distribution per size. Proposed dependencies are described by the linear equations with the correlation coefficients 0.97 and 0.84, respectively.

The metals, which are used by humanity in different fields of industrial production and technology, and different prefabricated metals are always corroded, since there are factors in the environment which accompany the corrosion processes, in particular, the oxygen of air, humidity, dust, etc. One among widely distributed methods of the metals and also the different prefabricated metals protection from the corrosion is an application of the anticorrosive coatings. Chapter 11 presents new methods of the nanosized inhibited pigments synthesis which have been developed based on the zinc phosphate and/or (poly)phosphate. This permits the creation of new competitive materials for their application in the paint and varnish industry under industrial conditions. It was shown that the acrylic monomers can be used as effective modifiers of the zinc phosphate nanoplates surface. Studying of the monomer nature and its concentration influence on a size and on a form of the obtained nanoplates, it was determined that at butyl methacrylate using as the surface's modifier the plates by the average size 200 ± 70 nm and by the thickness < 20 nm can be obtained.

There are a number of methods for the synthesis of different nanoparticles and nanomaterials; however, the kinetic peculiarities and regularities of the formation (nucleation and propagation) of nanoparticles are studied insufficiently. The aim of the studies presented in Chapter 12 was to investigate the kinetic regularities of the silver nanoparticles synthesis via reduction reaction of silver nitrate by hydrazine in the presence of sodium citrate depending on the concentration of the hydroxide ions and of the silver ions. As a result, the spherical silver nanoparticles (NPs) were obtained by silver ions reduction with hydrazine in the presence of sodium citrate as a stabilizer. The kinetic regularities of the silver NPs nucleuses formation and their propagation depending on the starting concentration of the hydroxide ions and silver ions were investigated. It investigated the influence of the synthesis conditions on the average diameter of the obtained silver NPs. It showed the dependence of the obtained silver NPs on kinetic parameters of the process.

In Chapter 13, several case studies on nanopolymers and their chemicals complexity are presented in detail. New approaches, along with their limitations and control, are discussed in detail.

This book will be useful first of all for scientists who are engaged in physical chemistry of polymers and their solutions as well as in nanochemistry and nanotechnologies. It will be helpful also for engineers and postgraduate training students interested in the progress of recent theoretical and practical advances in polymers and nanocomposites.

ABOUT AAP RESEARCH NOTES ON POLYMER ENGINEERING SCIENCE AND TECHNOLOGY

The AAP Research Notes on Polymer Engineering Science and Technology reports on research development in different fields for academic institutes and industrial sectors interested in polymer engineering science and technology. The main objective of this series is to report research progress in this rapidly growing field.

Gennady E. Zaikov, DSc
Head, Polymer Division, N. M. Emanuel Institute of Biochemical Physics, Russian Academy of Sciences; Professor, Moscow State Academy of Fine Chemical Technology, Russia; Professor, Kazan National Research Technological University, Kazan, Russia

---

Books in the AAP Research Notes on Polymer Engineering Science and Technology series
Functional Polymer Blends and Nanocomposites: A Practical Engineering Approach
Editors: Gennady E. Zaikov, DSc, Liliya I. Bazylak, PhD, and A. K. Haghi, PhD

Polymer Surfaces and Interfaces: Acid-Base Interactions and Adhesion In Polymer-Metal Systems
Irina A. Starostina, DSc, Oleg V. Stoyanov, DSc, and Rustam Ya.Deberdeev, DSc

Key Technologies in Polymer Chemistry
Editors: Nikolay D. Morozkin, DSc, Vadim P. Zakharov, DSc, and
Gennady E. Zaikov, DSc

Polymers and Polymeric Composites: Properties, Optimization, and Applications
Editors: Liliya I. Bazylak, PhD, Gennady E. Zaikov, DSc, and A. K. Haghi, PhD

CHAPTER 1

DISTRIBUTION OF INTERNAL LINKS OF THE POLYMER CHAIN IN THE SELF-AVOIDING RANDOMWALKS STATISTICS

YU. G. MEDVEDEVSKIKH

CONTENTS

SUMMARY

Within the frame of the self-avoiding random walks statistics (SARWS), the derivation of the internal n-link ($1 < n < N$) distribution of the polymer chain with respect to the chain ends is suggested. The analysis of the obtained expressions shows, that the structure of the conformational volume of the polymer chain is heterogeneous; the largest density of the number of links takes place in conformational volumes nearby the chain ends. It can create the effect of blockage of the active center of the growing macroradical and manifest itself as a linear chain termination. The equation for the most probable distance between two internal links of the polymer chain was obtained as well. The polymer chain sections, separated by fixing the internal links, are interactive subsystems. Their total conformational volume is smaller than the conformational volume of undeformed Flory coil. Therefore, total free energy of the chain sections conformation equals to free energy of the conformation of deformed (i.e., compressed down to the total volume of the chain sections) Flory coil.

1.1   INTRODUCTION

In *Gaussian* random walks statistics, the mean-square end-to-end distance R for a polymer chain, as well as mean-square distance between two not very closely located internal links obey general dependence [1]:

$$R = an^{1/2} \quad n > 1 \tag{1}$$

where a is the mean length of the chain link according to *Kuhn* [2]; n is the chain length or the length of a given chain section, expressed by the number of links in it.

Self-avoiding random walks statistics *(SARWS)* determines the conformational radius $R_{N,f}$ of the undeformed *Flory* coil as the most probable end-to-end distance of the polymer chain [3, 4]:

$$R_{N,f} = aN^{3/(d+2)} \tag{2}$$

where N is the total chain length, d is the *Euclidian* space dimension.

According to Eq. (2), *Flory* coil is a fractal, that is, an object, possessing the property of the scale invariance in dimensionality space $d_f = (d+2)/3$.

At derivation [3] of Eq. (2), however, the distribution of the internal polymer chain links in its conformational space remains unknown, therefore, it can not be indicated in advance that the distances between the terminal and internal chain links or between the internal ones obey the same dependence (Eq. 2) at the value of N as the length of the selected section of a polymer chain.

Study of the problem of internal polymer chain links' distribution is based mainly on the analysis [5, 6] of the scale distribution function $P_{ij}(r)$ of distance r between two links with ordinal numbers i and j:

$$P_{ij}(r) = |i - j|^{-d\upsilon} f\left(r / |i - j|^{\upsilon}\right)$$

(3)

Function $f\left(r / |i - j|^{\upsilon}\right) = f(x)$ is usually written in the form of power or exponential dependence on the only variable x:

$$f(x) \sim x^{\theta} \text{ at } x < 1,$$

$$f(x) \sim exp\{-x^{\delta}\} \text{ at } x > 1.$$

(4)

Studying the correlations between two arbitrary points i and j of a polymer chain, *Des Cloizeaux* [7] suggested dividing the scale function $P_{ij}(r)$ into three classes, that describe the distribution of distances between two terminal points of a polymer chain ($P^{(0)}_{ij}(r)$ with exponents θ_0 and δ_0 at $i = 1, j = N$), between the initial and internal points ($P^{(1)}_{ij}(r)$ with exponents θ_1 and δ_1 at $i = 1, 1 < j < N$) and between two internal points ($P^{(2)}_{ij}(r)$ with exponents θ_2 and δ_2 at $1 < i < j < N$), respectively.

Using the method of the second order ε-expansion within the range $x < 1$ for the space $d = 3$, *Des Cloizeaux* [7] has obtained in particular: $\theta_0 = 0.273$, $\theta_1 = 0.459$, $\theta_2 = 0.71$.

To evaluate the exponents θ_i and δ_i some other methods were used as well. Let us present some of the obtained results: [8] $\theta_0 = 0.27$; [8, 9, 10]

θ_1 = 0.55, 0.61, 0.70; [8, 9] θ_2 = 0.9, 0.67; [8, 11] δ_o = 2.44, 2.5; [8] δ_1 = 2.6; [8] δ_2 = 2.48.

In spite of the spread in exponent values, they unambiguously indicate (especially when comparing the values of θ_o, θ_1 and θ_2), that distribution function $P_{ij}(r)$, retaining their scale universality, quantitatively significantly depends on whether we consider the distance between terminal points, a terminal and internal one or between two internal points of a polymer chain. Whereas the proposed methods of analysis establish this fact, they however do not reveal the reason of the above-mentioned difference. Reference to strengthening the effects of the volume interaction between the internal links of a polymer chain can not be absolutized, since these effects can not be taken into account at computer simulation of self-avoiding random walks, but the results of the calculations according to them give the same estimations of exponents θ_i and δ_i as the analytic methods that take into account the volume interaction.

The shortcoming of the proposed approaches is also the fact that the scale distribution function $P_{ij}(r)$ is approximate and does not enclose the most significant region of parameter x changing between $x < 1$ and $x > 1$, where $P_{ij}(r)$ takes on maximal values. Finally, it should be noted that the role of the length of the second section of a polymer chain (at evaluating θ_1 and δ_1 the length of the second section is extrapolated to ∞) or the lengths of its two sections (at evaluating θ_2 and δ_2) is outside of the analysis.

Hence, the suggested approaches do not allow solving the problem of the internal links distribution for a polymer chain completely. In the present work we propose its analytic solution in terms of *SARW* strict statistics, that is, without taking into account of the so-called volume interaction.

1.2 INITIAL STATEMENTS

Preliminary let us briefly introduce the main statements of *SARW* statistics that are necessary for the subsequent analysis [3, 4]. The *Gaussian* random walks in N steps are described by the density of the *Bernoulli* distribution:

$$\omega(N,s) = \left(\frac{1}{2}\right)^N \prod_i \frac{n_i!}{[(n_i + s_i)/2]![(n_i - s_i)/2]!} \tag{5}$$

where n_i is a number of the random walk steps in i-direction of d-dimensional lattice space with the step length a, which is equal to the statistical length of *Kuhn* link; s_i is the number of effective steps in i-direction: $s_i = s_i^+ - s_i^-$, where s_i^+, s_i^- are numbers of positive and negative steps in i-direction. Numbers of n_i steps are limited by the following correlation:

$$\sum_i n_i = N \tag{6}$$

The condition of self-avoidance of a random walk trajectory on d-dimensional lattice demands the step not to fall twice into the same cell. From the point of view of chain link distribution over cells it means that every cell cannot contain more than one chain link. Chain links are inseparable. They cannot be torn off one from another and placed to cells in random order. Consequently, the numbering of chain links corresponding to wandering steps is their significant distinction. That is why the quantity of different variants of N distinctive chain links placement in Z identical cells under the condition that one cell cannot contain more than one chain link is equal to $Z!/(Z-N)!$

Considering the identity of cells, a priori probability that the given cell will be filled is equal to $1/Z$, and that it will not be filled is $(1-1/Z)$. Respectively, the probability $w(z)$ that N given cells will be filled and $Z-N$ cells will be empty, considering both the above mentioned condition of placement of N distinctive links in Z identical cells and the quantity of its realization variants will be determined by the following expression

$$\omega(Z) = \frac{Z!}{(Z-N)!}\left(\frac{1}{Z}\right)^N\left(1-\frac{1}{Z}\right)^{Z-N} \tag{7}$$

Probability density $\omega(N)$ of the fact that random walk trajectory is at the same time *SARW* statistics trajectory and at given Z, N, n_i will get the last step in one of the two equiprobable cells, which coordinates are set by vectors $s = (s_i)$, differentiated only by the signs of their components s_i, is equal to

$$\omega(N) = \omega(Z)\omega(N,s) \tag{8}$$

Let us find the asymptotic limit (Eq. 8) assuming $Z > 1$, $N > 1$, $n_i > 1$ under the condition $s_i < n_p$, $N < Z$. Using the approximated *Stirling* formula $\ln x!$ » $x \ln x - x + \ln (2p)^{1/2}$ for all $x > 1$ and expansion $ln(1-1/Z)$ » $-1/Z$, $ln(1-N/Z)$ » $-N/Z$, $ln(1 \pm s_i/n_i)$ » $\pm s_i/n_i - (s_i/n_i)^2/2$, and assuming also $N(N-1)$ we will obtain [3, 4]:

$$\omega(N) = \exp\{-N^2 / Z - (1/2)\sum_i s_i^2 / n_i\} \qquad (9)$$

Transition to the metric space can be realized by introduction of the displacement variable

$$x_i = a|s_i|d^{1/2} \qquad (10)$$

and also the parameter σ_i – the standard deviation of *Gaussian* part of distribution (9):

$$\sigma_i^2 = a^2 n_i d \qquad (11)$$

Then

$$s_i^2 / n_i = x_i^2 / \sigma_i^2, \qquad (12)$$

$$Z = \prod_i x_i / a^d \qquad (13)$$

and for the metric space Eq. (Eq. 9) becomes:

$$\omega(N) = \exp\left\{-\frac{a^d N^2}{\prod_i x_i} - \frac{1}{2}\sum_i \frac{x_i^2}{\sigma_i^2}\right\} \qquad (14)$$

Here $\prod_i x_i$ is the volume of conformational ellipsoid with the semi-axes of x_i, to the surface of which the states of the chain end belong.

A maximum of $\omega(N)$ at the set values of σ_i and N corresponds to the most probable, i.e., equilibrium state of the polymer chain. From the condition of $\omega(N)/\partial X_i = 0$ at $X_i = X_i^0$ we find semi-axes x_i^0 of the equilibrium conformational ellipsoid [3]:

$$x_i^0 = \sigma_i (a^d N^2 / \prod_i \sigma_i)^{1/(d+2)} \tag{15}$$

In the absence of external forces, all directions of random walks of the chain end are equiprobable, that allows to write:

$$n_i = N/d , \tag{16}$$

$$\sigma_i^2 = \sigma_N^2 = a^2 N . \tag{17}$$

The substitution of Eqs. (17) into (15) makes the semi-axes of the equilibrium conformational ellipsoid identical and equal to the undeformed *Flory* coil radius: $x_i^0 = R_{N,f}$. Let us underline two important circumstances. First, *SARW* statistics leads to the same result, that is, to Eq. (2), that *Flory* method, which takes into account the effect (repulsion) of the volume interaction between monomer links in the self-consistent field theory. However, as it was explained by *De Gennes* [6], accuracy of Eq. (2) in *Flory* method is provided by excellent cancellation of two mistakes: top-heavy value of repulsion energy as a result of neglecting of correlations and also top-heavy value of elastic energy, written for ideal polymer chain, that is in *Gaussian* statistics. Additionally, one must note, that Eq. (2) is only a special case of Eq. (15), which represents conformation of polymer chain in the form of ellipsoid with semi-axes $x_i^\circ \neq R_f$ allowing to consider this conformation as deformed state of *Flory* coil.

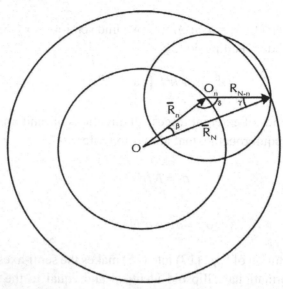

FIGURE 1   The scheme, explaining the necessity to enter new conformational volume Z'
at fixing of the position of the polymer chain internal n-link.

Second, obtained Eq. (14) for density of distribution of the end links of
polymer chain is not only more detailed but also more general than scale
dependencies (4) at θ_o and δ_o, which are approximately correct only at the
limits $x_i / R_i < 1$ and $x_i / R_f > 1$.

Free energy F_N of the equilibrium conformation is determined by the
expression

$$F_N = -kT \ln \omega(N) \text{ at } x_i = x_i^0 \tag{18}$$

From here for undeformed *Flory* coil we have:

$$F_{N,f} = (1 + d/2)kT(R_{N,f}/\sigma_N)^2. \tag{19}$$

For the deformed one-

$$F_N = F_{N,f}/\lambda_v, \tag{20}$$

where λ_v is the repetition factor of *Flory* coil's volume deformation:

$$\lambda_v = \prod_i x_i^o / R_f^d = \prod_i \lambda_i, \qquad (21)$$

where λ_i is a repetition factor of linear deformation,

$$\lambda_i = x_i^o / R_f. \qquad (22)$$

At any deformations the conformational volume diminishes, therefore in general case $\lambda_v \leq 1$ [3].

1.3   SARW STATISTICS FOR THE INTERNAL LINKS OF A CHAIN

The internal n-link ($1<n<N$) divides polymer chain into two sections with the lengths of n and $N-n$ links, respectively. This situation is illustrated with high quality in Fig. 1.

Let us assume, as it is shown in Fig. 1, that the most probable position of n-link with respect to the chain end is the surface of sphere with radius R_n and chain's n-link accidentally appeared at point O_n on this sphere. Then with respect to this point N-link of the chain over $N-n$ steps with the highest probability should appear on the surface of sphere with radius R_{N-n}.

As it can be seen from Fig. 1, $R_n^d + R_{Nnn}^d < R_N^d$ Consequently, fixing of n-link position diminishes the polymer chain's conformational volume regardless of where specifically point O_n is situated on the surface of sphere with radius R_n.

This means that for the analysis of *SARW* statistics of the chain's internal links in the lattice space, a new number of cells $Z' < Z$ needs to be introduced. Then the probability density of the random walk trajectory's self-avoiding for the polymer chain with fixed position of the internal link can be described by the *Bernoulli* distribution in the same form (Eq. 7), but with a new value of cells number:

$$\omega(Z') = \frac{Z'!}{(Z'-N)!}\left(\frac{1}{Z'}\right)^N\left(1 - \frac{1}{Z'}\right)^{Z'-N} \qquad (23)$$

The *Gaussian* random walks in n and $N-n$ steps of the first and second chain sections can be described thereby by the *Bernoulli* distribution (Eq.

5), but here in expressions for the distribution density $\omega(n, s)$ for the first chain section and $\omega(N-n, s)$ for the second one, respectively, the following conditions of normalization must be implemented:

$$\sum_i n_i = n \tag{24}$$

$$\sum_i n_i = N - n \tag{25}$$

and in place of the factor $(1/2)^N$ factors $(1/2)^n$ and $(1/2)^{N-n}$, respectively should be used.

As $\omega(Z')$ applies to the whole polymer chain, the distribution densities $\omega(n)$ and $\omega(N-n)$ of the *SARW* statistics trajectories for the first and second chain sections can be determined by the following expressions:

$$\omega(n) = (\omega(Z'))^{n/N} \omega(n,s), \tag{26}$$

$$\omega(N-n) = (\omega(Z'))^{(N-n)/N} \omega(N-n,s) \tag{27}$$

In an asymptotic limit the Eqs. (26) and (27) can be written:

$$\omega(n) = \exp\{-\alpha N^2/Z' - (1/2)\sum_i s_i^2/n_i\}, \tag{28}$$

$$\omega(N-n) = \exp\{-(1-\alpha)N^2/Z' - (1/2)\sum_j s_j^2/n_j\} \tag{29}$$

The lengths of every section fractions of the total chain length are introduced here as:

$$1 - \alpha = (N-n)/N \tag{30}$$

Defining the variables of the metric displacement of x_i and y_i in the form Eq. (10) and standard deviations $\sigma_{n,i}$ and $\sigma_{N-n,i}$ of the *Gaussian* part of distribution Eqs. (28) and (29) in the form Eq. (11), instead of Eqs. (28) and (29) we obtain:

$$\omega(n) = \exp\{-\alpha N^2 / Z' - (1/2)\sum_i x_i^2 / \sigma_{n,i}^2\},$$

(31)

$$\omega(N-n) = \exp\{-(1-\alpha)N^2 / Z' - (1/2)\sum_i y_i^2 / \sigma_{N-n,i}^2\}$$

(32)

Owing to the normalization Eqs. (24) and (25) we have:

$$\sum_i \sigma_{n,i}^2 = a^2 nd, \qquad \sum_i \sigma_{N-n,i}^2 = a^2(N-n)d$$

(33)

The values $\prod_i x_i$ and $\prod_i y_i$ are the volumes of the conformational ellipsoids with the semi-axes x_i and y_i of the first and second sections of the polymer chain, respectively. Hence, as laid down earlier Eq. (13), it is possible to write:

$$Z' = \left(\prod_i x_i + \prod_i y_i\right) / a^d.$$

(34)

Entering the volume fractions of the proper conformational ellipsoids

$$\beta = \prod_i x_i / \left(\prod_i x_i + \prod_i y_i\right), \quad 1 - \beta = \prod_i y_i / \left(\prod_i x_i + \prod_i y_i\right)$$

(35)

we obtain:

$$\omega(n) = \exp\{-a^d \alpha \beta N^2 / \prod_i x_i - (1/2)\sum_i x_i^2 / \sigma_{n,i}^2\},$$

(36)

$$\omega(N-n) = \exp\{-a^d (1-\alpha)(1-\beta)N^2 / \prod_i y_i - (1/2)\sum_i y_i^2 / \sigma_{N-n,i}^2\}$$

(37)

These expressions are sought densities of distribution of internal links of the chain from its ends. Parameter β will be determined later.

The most probable states of the polymer chain sections meet the conditions $\partial \omega(n)/\partial x_i = 0$ at $x_i = x_i^o$, $\partial \omega(N-n)/\partial y_i = 0$ at $y_i = y_i^o$. Using them and assuming that values β and $1-\beta$ do not depend on specific realizations of x_i and y_i, that is, these are functions of n and $N-n$ only, we find

$$x_i^0 = \sigma_{n,i}(a^d \alpha \beta N^2 / \prod_i \sigma_{n,i})^{1/(d+2)},$$

(38)

$$y_i^0 = \sigma_{N-n,i} (a^d (1-\alpha)(1-\beta) N^2 / \prod_i \sigma_{N-n,i})^{1/(d+2)} .$$ (39)

In the absence of external forces, all directions of random walks are equiprobable; therefore, according to Eq. (31) it is possible to write:

$$\sigma_{n,i}^2 = \sigma_n^2 = a^2 n = \sigma_N^2 \alpha$$ (40)

$$\sigma_{N-n,i}^2 = \sigma_{N-n}^2 = a^2 (N-n) = \sigma_N^2 (1-\alpha)$$ (41)

Using Eqs. (40) and (41) in (38) and (39), we will obtain expressions for the equilibrium conformational radii of both polymer chain sections:

$$R_n = R_{N,f} \alpha^{2/(d+2)} \beta^{1/(d+2)} ,$$ (42)

$$R_{N-n} = R_{N,f} (1-\alpha)^{2/(d+2)} (1-\beta)^{1/(d+2)} .$$ (43)

The conformational volumes here are equal to $\prod_i x_i = R_n^d$, $\prod_i y_i = R_{N-n}^d$, therefore Eq. (35) may be rewritten in the form

$$\beta / (1-\beta) = R_n^d / R_{N-n}^d$$ (44)

From (42) – (44) it follows:

$$\beta = \alpha^d / \left[\alpha^d + (1-\alpha)^d \right] \qquad 1-\beta = (1-\alpha)^d / \left[\alpha^d + (1-\alpha)^d \right]$$ (45)

Excepting β from Eqs. (42) and (43), we get finally

$$R_n = R_{N,f} \alpha / \left[\alpha^d + (1-\alpha)^d \right]^{1/(d+2)} ,$$ (46)

$$R_{N-n} = R_{N,f} (1-\alpha) / \left[\alpha^d + (1-\alpha)^d \right]^{1/(d+2)}$$ (47)

Eqs. (46) and (47) together determine the most probable, that is the equilibrium distances of the internal link from the polymer chain ends.

As one can see, although between R_n and R_{N-n} a simple correlation $R_n / R_{N-n} = \alpha / (1-\alpha)$ is observed, each of these values depends by complicated way not only on its own section length but also on the length of another one.

Eqs. (46) and (47) are correct at $d \leq 4$ [2], including at $d = 1$. For one-dimensional space from Eqs. (46) and (47) follows physically expected result $R_n = an$, $R_{N-n} = a(N - n)$.

1.4   STRUCTURE OF THE POLYMER CHAIN CONFORMATIONAL SPACE

From Eqs. (46) and (47) follows that $R_n / R_{N,f} \geq \alpha$ and $R_{N-n} / R_{N,f} \geq 1 - \alpha$, and signs of equality are achieved only on the chain ends, that is, at $\alpha = 1$ and $\alpha = 0$, respectively. This gives evidence to heterogeneity of the polymer chain conformational volume. In addition, because of interconnection between R_n and R_{N-n}, both chain sections are not fractals. Let us comment both circumstances, confronting the values R_n and R_{N-n} from Eqs. (46) and (47) with those values of $R_{n,f}$ and $R_{N-n,f}$, which these chain sections would have, if they were free and submitted to fractal correlation of the type (Eq. 2):

$$R_{n,f} = an^{3/(d+2)} = R_{N,f}\alpha^{3/(d+2)}, \tag{48}$$

$$R_{N-n,f} = a(N - n)^{3/(d+2)} = R_{N,f}(1 - \alpha)^{3/(d+2)} \tag{49}$$

From comparison of Eqs. (46)–(49) it follows:

$$R_n / R_{n,f} = \alpha^{(d-1)/(d+2)} / \left[\alpha^d + (1 - \alpha)^d \right]^{1/(d+2)} \tag{50}$$

$$R_{N-n} / R_{N-n,f} = (1 - \alpha)^{(d-1)/(d+2)} / \left[\alpha^d + (1 - \alpha)^d \right]^{1/(d+2)} \tag{51}$$

Dependencies of Eqs. (50) and (51) are illustrated on Fig. 2 for the option of $d = 3$.

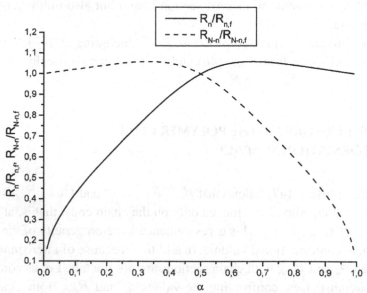

FIGURE 2 Ratios $R_n/R_{n,f}$ and $R_{N-n}/R_{N-n,f}$ calculated on the Eqs. (50) and (51) depending on a and $1-a$.

As one can see, only in the area $\alpha > 0.5$ and correspondingly $1-\alpha > 0.5$, ratios $R_n/R_{n,f}$ and $R_{N-n}/R_{N-n,f}$, though are more than 1, but insignificantly. It allows to consider of these chain sections as the fractals objects with a small error and to describe them by fractal dependences (48) and (49). However, for short chain sections, that is, at $\alpha < 0.5$ or $1-\alpha < 0.5$ ratios $R_n/R_{n,f}$ and $R_{N-n}/R_{N-n,f}$ become less than 1 and sharply diminish towards the chain ends, which indicates the compression of the conformational volume space nearby the ends of a chain.

Yet even more evidently heterogeneity of the structure of polymer chain's conformational volume becomes apparent at the analysis of volume density ρ, that is, the numbers of links in the unit of conformational volume for the given chain section. Let us be limited to considering only the first chain section n in length, for which

$$\rho_n = n / R_n^d \tag{52}$$

Using (46), we get

$$\rho_n / \rho_N = \left[\alpha^d + (1-\alpha)^d \right]^{d/(d+2)} / \alpha^{d-1}, \tag{53}$$

where $\rho_N = N / R_{N,f}^d$ is an average links' density in conformational volume of the whole polymer chain.

The correlation between local and average density of the chain links is illustrated on Fig. 3 at $d = 3$. Evidently, ratio ρ_n / ρ_N in the range of $\alpha < 0.2$ at $\alpha \to 0$ sharply increases, and for example at $\alpha = 0.01$ achieves the value of 10^4 order. As the dependence of the ratio $\rho_{N\ n} / \rho_N$ is similar, but asymmetric, it can be concluded that the conformational volumes near the chain ends are strongly compressed, so that the density of links in them considerably exceeds the average one over the conformational volume of the whole chain. With some caution one can suppose that the conformational volumes near the chain ends have a globular structure.

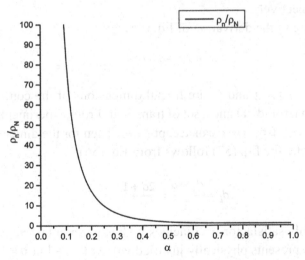

FIGURE 3   Ratio between local density of polymer chain link ρ_n and average one ρ_N depending on a calculated on Eq. (53).

To support this point, we propose also considerations based on experimental research of dymethacrylates post-polymerization kinetics, that is, dark, after turning off *UV* irradiation, process of polymerization [4]. It was found, that the chain termination is linear, and its kinetics submits to the law of stretched *Kohlrausch* exponent:

$$\psi(t) = g\exp\{-t/t_0\}^\gamma. \tag{54}$$

where g and $0 < \gamma \leq 1$ are constants; t_0 is characteristic time of linear chain termination.

A theoretical derivation [12] was based on the idea, that linear chain termination is the act of "self-burial" of macroradical's active center and manifests itself as the act of chain propagation, leading into a trap. Taking into account the fractal properties of polymer chain and assuming that a set of traps in its conformational volume is a fractal as well, we obtain the expression similar to Eq. (54):

$$\psi(t) = g\exp\left\{-t/\ \tau_p(\rho_m/\rho_0)^\xi\right\}^{1/\xi} \tag{55}$$

where τ_p is characteristic time of chain propagation act, ρ_m and ρ_0 are monomer and traps concentrations in the macroradical's conformational volume, respectively.

According to the derivation of Eq. (55)

$$\xi = d_f/(d_f + d_L - d), \tag{56}$$

where $d_f = (d+2)/3$ and d_L are fractal dimensions of the conformational volume of macroradical and a set of traps in it. From experimental data the value of $\gamma = 1/\xi = 0.6$, so we can accept $\xi = d_f$. Then for the dimension of the traps set fractal the Eq. (57) follows from Eq. (56):

$$d_L = \frac{d_f + d}{2} = \frac{2d+1}{3} \tag{57}$$

which not only satisfactorily coincides with experimental value of d_L at $d = 3$, but also presents physically justified value of $d_L = 1$ at $d = 1$.

Correlation of Eq. (57) shows that $d_f < d_L < d$. Therefore, in the reaction zone of growing macroradical, there are both "strange" traps formed by polymer chains, external for given macroradical, with fractal dimension close to d and "own" ones with fractal dimension of polymer chain $d_f = (d+2)/3$.

This derivation in Ref. [12] is based only on kinetic researches of dymethacrylates post-polymerization, but, apparently, it coordinates well with the results of present work, according to which "own" traps for a growing macroradical are caused by high density of links in the conformational volumes near the polymer chain ends, that can screen or even block up macroradical active center.

1.5 FREE ENERGY OF CONFORMATION OF POLYMER CHAIN SECTIONS

Let us determine free energies F_n and F_{N-n} of the conformation of polymer chain sections, separated by fixing of internal n-link, on type (Eq. 18) by expressions:

$$F_n = -kT \ln \omega(n) \text{ at } x_i = x_i^o \tag{58}$$

$$F_{N-n} = -kT \ln \omega(N-n) \text{ at } y_j = y_j^o. \tag{59}$$

For the equilibrium state in the absence of external forces, i.e., at all $x_i^o = R_n$ and $y_i^o = R_{N-n}$, using Eqs. (36) (37) (45)–(47) in Eqs. (58) and (59), we get

$$F_n = F_{N,f} \, \alpha \Big/ \Big[\alpha^d + (1-\alpha)^d \Big]^{2/(d+2)} \tag{60}$$

$$F_{N-n} = F_{N,f} (1-\alpha) \Big/ \Big[\alpha^d + (1-\alpha)^d \Big]^{2/(d+2)} \tag{61}$$

From here it follows:

$$F_n + F_{N-n} = F_{N,f} \Big/ \Big[\alpha^d + (1-\alpha)^d \Big]^{2/(d+2)} \tag{62}$$

Hence, $F_n + F_{Nnn} n \geq F_{N,f}$. It is related to the fact that two chain sections are thermodynamic subsystems, which interact with each other. Thus, fixing the position of a polymer chain internal link increases its negative entropy and positive free energy of conformation due to diminishment of the polymer chain conformational volume. Therefore the sum $F_n + F_{N-n}$ of free energies of the chain sections conformation must be compared not to $F_{N,f}$, representing free energy of undeformed *Flory* coil conformation with the volume of $R_{N,f}^d$, whereas for free energy of conformation F_N of the *Flory* coil deformed to the volume $R_n^d + R_{N-n}^d$, determined by the Eq. (20).

As the multiple (repetition factor) of the volume deformation λ_V here is equal to

$$\lambda_V = \left(R_n^d + R_{N-n}^d \right) \big/ R_{N,f}^d = \left[\alpha^d + (1-\alpha)^d \right]^{2/(d+2)} \tag{63}$$

we have identically

$$F_n + F_{N-n} = F_{N,f} \big/ \lambda_V \tag{64}$$

It is now possible to accomplish the reverse transition and write the expression

$$\left[\omega(n)\omega(N-n) \right]^{\lambda_V} = \omega(N), \tag{65}$$

which was not obvious in the beginning.

1.6 THE MOST PROBABLE DISTANCE BETWEEN TWO INTERNAL LINKS OF A POLYMER CHAIN

If two internal links of a chain are selected according to the condition $1 < k < n < N$, the polymer chain is divided into three sections with the lengths of k, n-k and N-n, to which the fractions of the total chain length $\alpha_k = k/N$, $\alpha_{n-k} = (n-k)/N$ correspond.

Let us suggest that $j = \overline{1,3}$ are numbers of sections with quantities of links k, $n - k$ and $N - n$, to which fractions $\alpha_1 = k/N$, $\alpha_2 = (n-k)/N$ and $\alpha_3 = (n-k)/N$ from general number N of links in a chain are corresponded.

Extending the above-mentioned procedure of analysis of two chain sections to three sections, we get the general expression for the distribution density of the end of the given section regarding to its beginning:

$$\omega(j) = \exp\{-a^d \alpha_j \beta_j N^2 / \prod_i x_{ji}^2 - (1/2)\sum_i x_{ji}^2 / \sigma_{ji}^2\} \tag{66}$$

where: β_j is a fraction of conformational volume of the given section in the sum of conformational volumes of all sections; x_{ji} $i = \overline{1,d}$ are semi-axes of conformational ellipsoid j with center in the beginning of the given section. The surfaces of this section involve the states of its end.

The square deviations σ_{ji} of *Gaussian* part of distribution (Eq. 66) obey the normalization conditions of the form:

$$\sum_i \sigma_{ij}^2 = a^2 \alpha_j Nd \tag{67}$$

At equiprobability of walks in all directions of d-dimensional space we have:

$$\sigma_{ij}^2 = \sigma_{j0}^2 = a^2 \alpha_j N \tag{68}$$

In this case the most probable distance between the beginning and the end of the chain in the given section will be equal to:

$$R_j = \frac{R_{N,f}\alpha_j}{\sum_j (\alpha_j^{d})^{1/(d+2)}} \tag{69}$$

and for β_j the following expression will be correct:

$$\beta_j = \frac{\alpha_j^d}{\sum_i \alpha_j^d} \tag{70}$$

According to Eq. (69) the value of R_j depends not only on the length of the given chain section but also on where this section has been chosen. Thus, we can note again that selected three sections of polymer chain are not independent, but are interactive subsystems. Therefore, total free energy of conformation of the chain three sections exceeds free energy of conformation of undeformed Flory coil. But the following equality holds identically:

$$F_k + F_{n-k} + F_{N-k} = F_{N,f}/\lambda_v , \tag{71}$$

where the multiplicity of *Flory* coil's volume deformation at the division of the chain into three sections is determined by the expression:

$$\lambda_v = \left(R_k^d + R_{n-k}^d + R_{N-n}^d \right) \Big/ R_{N,f}^d \tag{72}$$

1.7 CONCLUSIONS

Fixation of the position of polymer chain internal links separates its conformational volume into interacting subsystems. Their total conformational volume is smaller, and free energy is larger than the conformational volume and free energy of *Flory* coil, respectively. From expressions, which determine the probable distance between the polymer chain's internal link and its ends, as well as between any internal links, it follows that the structure of the polymer chain conformational volume is heterogeneous: the largest density of the number of links is observed near the chain ends. This can result in blockage of the macroradical's active center and appear as a linear chain termination.

KEYWORDS

- **conformation**
- **distribution**
- **internal link**
- **macromolecule**
- **solution**

REFERENCES

1. Grossberg, Yu. A., Khokhlov, A. R. Statistical Physics of Macromolecules, *Science Edition*, Moscow, 344 p. (1989) *(in Russian)*.
2. Kuhn, W. *Polymers*, **76**, 258 (1936).

3. Medvedevskikh, Yu. G. *Condensed Matter Physics*, **4** (2), 209–219 (2001).
4. Medvedevskikh, Yu. G. *J. Appl. Polym. Sci.,* **109**, 2472–2481 (2008).
5. Des Cloizeaux, J. *Phys. Rev. A.*, **10**, 1665 (1974).
6. De Gennes, P. G. Scaling Concepts in Polymer Physics, *Cornell University Press, Ithaka-London*, 368 p. (1979).
7. Des Cloizeaux, J. *J. Phys. Paris*, **41**, 223 (1980).
8. Baumgartner, A. *Z. Phys.*, **42**, 265 (1981).
9. Guttman, A. J., Sykes, M. F. *J. Phys.*, C **6**, 945 (1973).
10. Whittington, S. G., Trueman, R. E., Wilker, J. B. *J. Phys.*, A **8**, 56 (1975).
11. Redner, S. *J. Phys.*, A **13**, 3525 (1980).
12. Medvedevskikh, Yu. G., Kytsya, A. R., Bazylyak, L. I., et al., In: *Conformation of Macromolecules: Thermodynamic and Kinetic Demonstrations* Edited by Yu. G. Medvedevskikh, S. A. Voronov, G. E. Zaikov *(Nova Science Publishers, New York)*, 173–192 (2007).

CHAPTER 2

POLYMERIC STARS IN SELF-AVOIDING RANDOMWALKS STATISTICS

YU. G. MEDVEDEVSKIKH, L. I. BAZYLYAK, and G. E. ZAIKOV

CONTENTS

SUMMARY

It was proposed the self-avoiding random walks statistics for polymeric similar starts in diluted and concentrated solutions. On a basis of the proposed statistics the equilibrium thermodynamical properties (namely, volume, pressure, free energy) of the conformational state of polymeric stars and their transformation under the action of external forces have been described. It were determined the elastic properties (Yung's modulus and modulus of shearing) of the conformational volume of polymeric starts and also obtained the expressions for the main forces and for the deformation work. The dynamical properties of polymeric starts, in particular their characteristic times of translational and rotational motions, on a basis of which the coefficients of diffusion and elastic component of a viscosity determined have been studied. In concentrated solutions all the thermo-dynamic and dynamical properties expressed not only as the functions on a length and on a number of rays, but also on the concentration of the polymeric stars.

2.1 INTRODUCTION

Polymeric stars attract the attention not only due to the own technological properties [1–3], which can be varied by a length N of the rays and by their number s, but also as the complicated and interesting object of the statistical physics of polymers.

Experimental data concerning to the properties of the polymeric stars are not numerous [4–7] and doesn't give the sufficient well-defined picture. In particular, an investigation of the low–angle scattering of the neutrons [5, 6] for the diluted solutions of the star-like polybutadienes with 3, 4, 8 and 12 rays, one among which is deuterated, points on the fact, that the gyration radius of the tagged ray is increased at the number of the rays growing, but at this in diluted solutions the gyration radius is more, than in a melt.

Rheological investigations [7] of the regular polystirene starts which are differed by a number of the rays (6, 12 and 22) and by the structure of the branching center (one or two covalently connected molecules of fullerene) also give the contradictory results: the characteristic viscosity of their solutions in chloroform is weakly increased at the molar mass of the star growing (with index $\alpha = 0.2$ in *MKH* equation $[\eta] = M^{\alpha}$), whereas it is decreased in tetrahydrofurane, in other words, has the negative index $\alpha = -0.06$.

Theoretical analysis of the polymeric stars in *Daud-Cotton's* model [8], which considered them as spherical polymeric brushes in quasi–plate approximation [9–12] leads to the dependence of the conformational radius R_s of polymeric star on a number of rays s and their length N in a form (d - is dimensionality of a space):

$$R_s = aN^{3/(d+2)}s^{1/(d+2)}, \tag{1}$$

The Eq. (1) has an essential lack: at $s = 2$, that is, under condition of the polymeric star degeneracy into the linear chain by $2N$ length, it wrongly illustrates of this situation.

Besides, the experimental value of the so-called branching index g_R, determining by ratio

$$g_R = R_s^2 \big/ R_{fs}^2, \tag{2}$$

where R_{fs} is the conformational radius of the linear polymeric chain containing as same as the star of sN links,

$$R_{fs} = a(sN)^{3/(d+2)}, \tag{3}$$

is disagreed with the theoretical values g_R, calculated accordingly to Eqs. (1)-(3).

In statistical physics like to the linear chains for polymeric stars two main tasks are solved. First is the estimation of a number of allowed configurations L with the calculation of indexes γ and γ_s in the scaling expressions by following type [13]:

$$L \approx z^N N^{\gamma-1} \tag{4}$$

for linear chain and

$$L_s \approx z^{sN} N^{\gamma_s - 1} \tag{5}$$

for the star.

In Eqs. (4) and (5) z is not strictly determined; at the scaling approach the effective coordination number of d–measured lattice, parameters γ and γ_s are assumed as universal scaling indexes.

For linear chain at $d = 3$ the value $\gamma = 1.1596 \pm 0.002$ was obtained [14] using the method of a group of the field theory renormalization. Index γ_s for polymeric star calculating by several methods including the groups of the field theory renormalization [15–18], methods of molecular dynamics [19–20] and *Monte-Carlo's* method [21–23] is sharply decreased at the s growing, taking more negative values even at $s > 8$ (for example, at $s = 32$ an estimation with the use of *Monte-Carlo's* method [23] gives the value $\gamma_s = -29$); this prejudices the universality of this scaling index.

Theoretical calculation of the L and L_s in statistics of the self-avoiding random walks *(SARW)* is represented in Chapter 4 of this collection book. Here let's consider the second main task, which consists in the estimation of the distribution function of the end of a chain from its beginning.

More often the distribution function $P(N)$ of the distance r between the ends of the linear chain by the N length is written in the scaling form [24–25]:

$$P(N) = R_f^{-d} \varphi(r / R_f), \tag{6}$$

In which R_f is the conformational radius of *Flory*

$$R_f = aN^{3/(d+2)} \tag{7}$$

Function $\varphi(r / R_f)$ represented by two asymptotes:

$$\varphi(x) \approx x^{\theta} \quad \text{at } x < 1 \tag{8}$$

$$\varphi(x) \approx \exp\left\{ -x^{\delta} \right\} \text{ at } x > 1 \tag{9}$$

with indexes, for example, $\theta = 0.273$ [26] and $\delta = 2.5$ [27].

The same scaling ratios of Eqs. (6) (8) and (9) are used for the polymeric starts, but the indexes θ and δ are the functions on the number of rays.

Let's note, finally, that the scaling approach to the description of the distribution function in a form (Eq. 6) in spite of its universality is approximate and limited. In particular, it not covers the most important field of the parameter x in (8) and (9) between $x < 1$ and $x > 1$ change, in which $P(N)$ takes the maximal values corresponding to the most probable conformational state of the polymeric chain. That is why, even the calculation of the indexes θ and δ doesn't give the possibility to describe the thermodynamical properties of the conformational state of polymeric chains and their transformation, for example at the deformation; this is not allow strictly to estimate the elastic properties of the conformational volume.

Exact form of the distribution function and following from it thermodynamical properties of linear polymeric chain conformation are strictly determined in the *SARW* statistics for ideal diluted [28] and concentrated [29] solutions. Here this approach is spread on the regular polymeric stars in diluted and concentrated solutions with the description of their thermodynamical and dynamical properties.

2.2   DILUTED SOLUTION OF POLYMERIC STARS

2.2.1   *SELF-AVOIDING RANDOM WALK STATISTICS*

Let polymeric star with the s rays by the same length N is inserted into d-measured lattice space with the parameter of a cubic cell equal to the statistical length of a monomeric link of the rays of star, and let Z is the number of cells of the lattice space, in which there are the all sN links of a star. We will be neglect by the effects of interactions, assuming that all the possible configurations of polymeric star are energetically equal.

Let mark any undefined pair of rays and will be considered it as the linear polymeric chain by length of $2N$. Let fix its one end as the referenced position of the Gaussian phantom walks of the second end. Let define via

n_i steps of walks the end of the marked chain relatively it's beginning along the i directions of d-measured lattice space, limited by the condition of the normalization

$$\sum_i n_i = 2N \tag{10}$$

The number of methods for the realization of walks along the i direction is equal to $n_i!/n_i^+!n_i^-!$, where the numbers of the steps into positive n_i^+ and into negative n_i^- directions connected via the ratio $n_i^+ + n_i^- = n_i$. Since the a priori probability of a choice of the positive or negative direction for every step is the same and is equal to $1/2$, the probability of $\omega(n_i)$ that under given n_i will be done of n_i^+ positive and n_i^- negative steps, will be determined by the *Bernoulli's* equation:

$$\omega(n_i) = \left(\frac{1}{2}\right)^{n_i} n_i!/n_i^+!n_i^-! \tag{11}$$

Inputting the numbers of the resulting steps via the ratio $s_i = n_i^+ + n_i^-$, we have $n_i^+ = (n_i + s_i)/2$, $n_i^- = (n_i - s_i)/2$. Then the Eq. (11) can be rewritten as follow:

$$\omega(n_i) = (1/2)^{n_i} n_i!/((n_i + s_i)/2)!((n_i - s_i)/2)! \tag{12}$$

For the walks along the all directions of the d-measured space we have

$$\omega(n) = (1/2)^N \prod_i n_i!/((n_i + s_i)/2)!((n_i - s_i)/2)! \tag{13}$$

The condition of the absence of self-intersection trajectories of walks requires from the point of view of the links of star per cells distribution that in every cell of the lattice space should be exist not more than one link of a star. The links of a polymeric chain are continuous; they cannot be separated one from other and located upon the cells into the undefined order. Therefore, the number of a links in a chain is its essential distinctive feature. The links of the different rays are also distinguishable. That is why the number of different methods location of sN distinguishable links of a star per Z similar cells under condition that in every cell cannot be more than one link is equal to $Z!/(Z-sN)!$. Due to the identity of the cells, a priori probability of fact that the presented cell will be occupied by a link is equal to $1/Z$, will be not occupied - $(1-1/Z)$. Therefore, the probability $\omega(z)$ of sN distribution of the distinguishable links of a star per Z identical cells under condition, that in every cell can be not more than one link, will be determined by the *Bernoulli's* distribution:

$$\omega(z) = \frac{Z!}{(Z-sN)!}(\frac{1}{Z})^{sN}(1-\frac{1}{Z})^{Z-sN} \tag{14}$$

Distribution Eq. (13) describes the trajectory of the walks of undefined marked linear chain by the length $2N$, whereas the Eq. (14) determines the distribution of the all sN links of a star upon the Z identical cells. That is why the probability $\omega(2N)$ of the common fact consisting in that the trajectory of the random walk for the chosen chain by the $2N$ length is the trajectory without the self-avoiding is equal

$$\omega(2N) = (\omega(Z)^{2/s}\omega(n) \tag{15}$$

Here the index $2/s$ is the result of the division of $2N$ on sN,

Using the *Stirling's* formula by the following type $\ln N! \cong N \ln N - N$ at $N > 1$, $Z \gg sN$, $N > 1$, $n_i > 1$ and decomposition $\ln(1-1/Z) \cong -1/Z$, $\ln(1-sN/2Z) \cong -sN/2Z$, $\ln(1 \pm s_i/n_i) \cong \pm s_i/n_i - (s_i/n_i)^2/2$ in accordance with the condition $s_i < n_i$, and assuming $N(N-1) = N^2$, we will obtained the asymptotic form (Eq. 15) with the accuracy to the constant multiplier neared to the one:

$$\omega(2N) = \exp\left\{-\frac{(s/2)(2N)^2}{Z} - \frac{1}{2}\sum_i \frac{s_i^2}{n_i}\right\} \tag{16}$$

The transition to the metric d-measured space is possible via the introduction of the variable of the displacement x_i along the i direction of the end of a chain from its beginning,

$$x_i = a|s_i|d^{1/2}, \tag{17}$$

and also of the standard deviation σ_i of the Gaussian part in Eq. (16)

$$\sigma_i^2 = a^2 n_i d \qquad (18)$$

Accordingly to the condition of the normalization (10) the following relationship is superposed on the value σ_i:

$$\sum_i \sigma_i^2 = a^2 2Nd \qquad (19)$$

The Eq. (16) is true for any undefined chosen pair of the rays representing by itself the linear chain by the length $2N$. Therefore, for the all of them Z is general parameter characterizing in metric space the conformational volume of polymeric star:

$$Z = \prod_i x_i \Big/ a^d \qquad (20)$$

So, taking into account the Eqs. (17)-(20), a distribution (Eq. 16) can be rewritten as follow:

$$\omega(2N) = \exp\left\{ -a^d (s/2)(2N)^2 / \prod_i x_i - (1/2)\sum_i x_i^2 / \sigma_i^2 \right\} \qquad (21)$$

Distribution (Eq. 21) in metric space determines the probability $\omega(2N)\prod_i dx_i$ of fact that the trajectory of the self-avoiding walk of any undefined pair of rays representing by itself the linear chain by $2N$ length, under the given values of s, N and σ_i will be finished into elementary volume $\prod_i dx_i$, locating on the surface of ellipsoid with the semi-axes x_j, $i = 1,d$.

2.2.2 THERMODYNAMICS OF CONFORMATION AND DEFORMATION OF POLYMERIC STARS

Without taking into account of the energetic effects of the interactions, the main factor determining the thermodynamic properties of the conformational state of polymeric star is the entropy of the self–avoiding random walks, which accordingly to the *Boltzmann* is determined by the thermodynamical probability of a state, that is by the number of methods of its realization; such methods are the trajectories of the self–avoiding random walks or configurations.

A connectedness of the links in polymeric chain imposes the first and essential limitation on the trajectories of walks – an inhibition of the step backwards. Consequently, the number of variants of a step of walk trajectory cannot be more than $2d-1$. Only the first step has $2d$ variants of the motion. This means, that the maximal number L_{max} of the trajectories of walk for undefined pair of rays will be equal

$$L_{max} = (2d / (2d-1))(2d-1)^{2N} \tag{22}$$

If the presented state is realized by L methods, then its probability is equal to L / L_{max}, that is why let determine the number $L(2N)$ of configurations for the undefined chosen pair of the rays via expression:

$$\omega(2N) = L(2N) / L_{max} \tag{23}$$

It's follow from this:

$$L(2N) = (2d / (2d-1))(2d-1)^{2N} \omega(2N) \tag{24}$$

For polymeric star the number of independent pairs from s rays is equal to $s(s-1)/2$. That is why, the number of methods $L(sN)$ for the realization of conformational state of polymeric star will be equal:

$$L(sN) = \left(L(2N)\right)^{s(s-1)/2} \tag{25}$$

Therefore, in accordance with *Boltzmann*, the general entropy $S_0 = k \ln L(sN)$ of the conformational state of polymeric star can be represented as two components:

$$S_0 = k \frac{s(s-1)}{2} \ln \frac{2d}{2d-1} (2d-1)^{2N} + k \frac{s(s-1)}{2} \ln \omega(2N) \tag{26}$$

The first term is positive and is more than the second one; it takes into account the all trajectories of walk with imposed on them singular limitation of the connectedness of the links into a chain, and doesn't accept the reverse step. The second term is negative ($\omega(2N) < 1$); it takes into account the additional limitations on the trajectories of walk by requirement of their self-intersection absence. At this, the first term at given data s, N, d is the constant value, whereas the second term via $\omega(2N)$ is the function on the conformational state. That is why let assume only the second term in Eq. (26) as the entropy S of the conformational state of a polymeric star:

$$S = k(s(s-1)/2)\ln \omega(2N) \tag{27}$$

Negative value of determined in such a way entropy of the conformation permits to consider the conformation as the statistical form of the self-organization of polymeric stars, and the numerical measure of this self-organization is entropy accordingly to Eq. (27).

Free energy of the conformational state without taking into account of the energetic effects is equal to $F(x) = -TS$. By combining of Eqs. (21) and (27), we will obtain:

$$F(x) = kT \frac{s(s-1)}{2}\left[\frac{a^d(s/2)(2N)^2}{\prod_i x_i} + \frac{1}{2}\sum_i \frac{x_i^2}{\sigma_i^2}\right] \tag{28}$$

Next, among all the possible conformational states let mark the most probable or thermodynamically equilibrium conformational states, to which the maximum S and the minimum F correspond in accordance with the condition $\partial F(x)/\partial x_i = 0$ at $x_i = X_i$. By differing of Eq. (28), we have

$$\partial F(x)/\partial x_i = kT \frac{s(s-1)}{2}\left[-\frac{a^d(s/2)(2N)^2}{x_i \prod_i x_i} + \frac{x_i}{\sigma_i^2}\right] \tag{29}$$

By equaling of the right part to zero and solving the system consisting of the algebraic equations for the all i, we will find the most probable or equilibrium values of X_i, which are the semi-axes of the equilibrium conformational ellipsoid:

$$X_i = \sigma_i \left(a^d(s/2)(2N)^2 / \prod_i \sigma_i\right)^{1/(d+2)} \tag{30}$$

Under the absence of the external forces and into the ideal solution the all directions of the walk are equiprobable accordingly to condition $n_i = 2N/d$;

that is why parameters σ_i in accordance with the Eq. (18) take the same value equal to:

$$\sigma_0^2 = a^2 2N \qquad (31)$$

The substitution of values $\sigma_i = \sigma_0$ in (30) based on Eq. (31) makes the semi-axes of the equilibrium conformational ellipsoid the same and equal to the conformational radius of the polymeric star:

$$R_s = a(2N)^{3/(d+2)}(s/2)^{1/(d+2)} \qquad (32)$$

Let's estimate the branching index g_R determined by the ratio (Eq. 2) with the use of the Eq. (3) for R_{fs} and Eq. (32) for R_s:

$$g_R = (s/2)^{-4/(d+2)} \qquad (33)$$

Calculations accordingly to Eq. (33) for variant $d = 3$ at $s = 3$ and $s = 12$ give the values equal to 0.73 and 0.24 respectively, which are good agreed with the experimental data 0.78 and 0.24 [31].

Under the action of external forces along the axes of d-measured space appearing in particular at the transition of polymeric star from the ideal solution into the real one, $\sigma_i \neq \sigma_0$ and spherical conformational space of the polymeric star is deformed into the ellipsoid with the semi-axes X_i accordingly to Eq. (30), equilibrium as to σ_i. We assume the following variables as a measure of the conformational volume deformation:

$$\lambda_i = X_i / R_s \qquad (34)$$

$$\lambda_v = \prod_i X_i / R_s^d = \prod_i \lambda_i, \qquad (35)$$

which represent by themselves the multiplicities of linear and volumetric deformation respectively.

In accordance with the condition of normalization (Eq. 10) the values of λ_i cannot accept the undefined values. Let's determine the relationship between them by introducing the secondary parameters:

$$\varphi_i = \sigma_i / \sigma_0 \tag{36}$$

It is follows from the determinations of σ_i from Eq. (18) and of σ_0 from Eq. (31):

$$\sum_i \phi_i^2 = d \tag{37}$$

Bu substitution of the values $\sigma_i = \varphi_i \sigma_0$ in Eq. (30) and taking into account (Eq. 34) we will find

$$\lambda_i = \phi_i \Big/ \left(\prod_i \phi_i \right)^{1/(d+2)} \tag{38}$$

Therefore,

$$\lambda_v = \prod_i \lambda_i = \left(\prod_i \phi_i \right)^{2/(d+2)}, \tag{39}$$

so

$$\phi_i = \lambda_i \lambda_v^{1/2} \tag{40}$$

Accordingly to Eqs. (37) and (40) we have

$$\sum_i \lambda_i^2 = d / \lambda_v \tag{41}$$

An analysis of Eq. (41) shows, that under any deformation of the spherical conformational volume of polymeric star, to which the values of $\lambda_i = 1, i = 1.d$ and $\lambda_v = 1$ correspond, into the conformational ellipsoid, the multiplicity of the volumetric deformation is decreased ($\lambda_v \leq 1$);, that means the compaction of the conformational volume of polymeric star.

Next, let's determine a free energy $F(\lambda)$ of the equilibrium conformational ellipsoid by substitution in Eq. (28) of values $x_i = X_i = R_s \lambda_i$:

$$F(\lambda) = kT \frac{d+2}{2} \left(\frac{R_s}{\sigma_0} \right)^2 \frac{s(s-1)}{2} / \lambda_v \tag{42}$$

Here

$$(R_s / \sigma_0)^2 = (2N)^{(4-d)/(d+2)} (s/2)^{2/(d+2)} \tag{43}$$

That is why it can be also written:

$$F(\lambda) = kT((d+2)/2)(2N)^{(4-d)/(d+2)} (s/2)^{(d+4)/(d+2)} (s-1)/\lambda_v \tag{44}$$

For strainless state $\lambda_v = 1$, that is why a work of its transition into a deformed state in the system of the mechanics signs will be equal:

$$\Delta F_{def} = kT \frac{d+2}{2} \left(\frac{R_s}{\sigma_0} \right)^2 \frac{s(s-1)}{2} \left(\frac{1}{\lambda_v} - 1 \right) \tag{45}$$

Such work is positive, that is in the system of the mechanics signs it's realized under the system. If, however, the conformational volume is changed from the one deformed state with λ_v into another deformed one with λ_v'', then the deformation work will be equal to:

$$\Delta F_{def} = kT \frac{d+2}{2} \left(\frac{R_s}{\sigma_0} \right)^2 \frac{s(s-1)}{2} \left(\frac{1}{\lambda_v''} - \frac{1}{\lambda_v'} \right) \tag{46}$$

and can be characterized by any sign.

Let's determine the pressure of conformation P of polymeric star via usual thermodynamic expression

$$\partial F(\lambda)/\partial V = -P \tag{47}$$

as a measure of relationship between free energy and volume of conformation. Since the conformational volume is equal to $V = R_s^d \lambda_v$ we have

$$P = -(\partial F(\lambda)/\partial \lambda_v)/R_s^d \tag{48}$$

By differing the Eqs. (42) and (44) upon λ_v, respectively, we will obtain:

$$P = kT \frac{d+2}{2} \left(\frac{R_s}{\sigma_0} \right)^2 \frac{s(s-1)}{2} \bigg/ R_s^d \lambda_v^2 , \tag{49}$$

$$P = \frac{kT}{a^d}\frac{d+2}{2}(2N)^{4(1-d)/(d+2)}(s/2)^{4/(d+2)}(s-1)/\lambda_v^2 \qquad (50)$$

It's follows from the comparison of Eqs. (42) and (49), that the pressure of conformation numerically is equal to density of free energy: $P = F(\lambda)/V$. Next, by multiplying of this expression on $V^2 = \left(R_s^d \lambda_v\right)^2$ and taking into account the Eqs. (42) and (49), we will find the equation for the conformation state of the polymeric star in a form

$$PV^2 = FV = const \qquad (51)$$

where

$$const = kT\frac{d+2}{2}\left(\frac{R_s}{\sigma_0}\right)^2 \frac{s(s-1)}{2} R_s^d \qquad (52)$$

So, the values $PV^2 = FV$ are integrals of the process of equilibrium deformation of the conformational volume of polymeric star.

2.2.3 MODULUS OF ELASTICITY

Under approximation of the isotropy of conformational volume of polymeric star its relative deformation $\partial x_i / x_i$ in i direction of d-measured space under the action of the all main forces f_i let's express via the differential form of the *Poisson's* equation:

$$Y\frac{\partial x_i}{x_i} = \frac{\partial f_i}{\prod_{j\neq i} x_j} + \gamma\sum_{j\neq i}\frac{\partial f_j}{\prod_{k\neq j} x_k} \qquad (53)$$

where: Y is the *Young's* modulus, γ is the *Poisson's* coefficient, $Y\partial x_i / x_i$ is the tension in $(d-1)$-measured plate normal to the i direction; $\prod_{j\neq i} x_j$ and $\prod_{k\neq j} x_k$ are values of the areas which are normal to the forces f_i and f_j, respectively.

Let's rewrite Eq. (53) to the *Young's* modulus:

$$Y = \frac{x_i^2}{\prod_i x_i} \frac{\partial f_i}{\partial x_i} + \gamma \sum_{j \neq i} \frac{x_i x_j}{\prod_i x_i} \frac{\partial f_j}{\partial x_i} \tag{54}$$

In the system of the mechanics signs $f_i = \partial F(x)/\partial x_i$, that permits to use the Eq. (29). Under equilibrium values $x_i = X_i$ these forces (but not their derivatives) are equal to zero. That is why by differing of Eq. (29) upon x_i and x_j, and next by substituting of the equilibrium values $x_i = X_i = R_s \lambda_i$ we will obtain

$$\partial f_i / \partial x_i = 3kTs(s-1)/2\sigma_0^2 \lambda_i^2 \lambda_v , \tag{55}$$

$$\partial f_i / \partial x_j = \partial f_j / \partial x_i = kTs(s-1)/2\sigma_0^2 \lambda_i \lambda_j \lambda_v \tag{56}$$

By substituting of these derivatives in Eq. (54) with change of x_i on the equilibrium values X_i we will find

$$Y = kT(3 + \gamma(d-1))(R_s/\sigma_0)^2 \frac{s(s-1)}{2} \Big/ R_s^d \lambda_v^2 \tag{57}$$

Comparing the Eqs. (57) and (49), we find the relationship between the *Young's* modulus and the pressure of conformation:

$$Y = 2(3 + \gamma(d-1))P/(d+2) \tag{58}$$

From the other hand, in general case of the d-measured space the relationship between Y and P can be expressed via the volumetric modulus

$$E = -VdP/dV \tag{59}$$

by the ratio

$$E = Y/d(1 - \gamma(d-1)) \tag{60}$$

It's follows from the determination of Eq. (59) and the equation of state (Eq. 51):

$$E = 2P \tag{61}$$

Substituting of Eqs. (61) into (60) we will obtain another equation of relationship between Y and P:

$$Y = 2d(1 - \gamma(d-1))\ P \qquad (62)$$

Comparing the Eqs. (62) and (58) we find the expression for the *Poisson's* coefficient:

$$\gamma = (d+3)/(d+1)^2 \qquad (63)$$

Next, we determine the shear modulus via the *Young's* modulus and the *Poisson's* coefficient for $d \geq 2$-measured space

$$\mu = Y/(d-1)(1+\gamma) \qquad (64)$$

which is also the function on the pressure of conformation:

$$\mu = \frac{2(3 + \gamma(d-1))}{(d+2)(d-1)(1+\gamma)} P \qquad (65)$$

2.2.4 MAIN FORCES AND THE WORK OF DEFORMATION

It follows from the determination of the main forces $f_i = \partial F(x)/\partial x_i$ in accordance with the Eq. (29) that at the equilibrium values $x_i = X_i$ these forces are equal to zero. That is why let's determine the main forces as those, which should be applied to the strainless conformational volume R_s^d, for which the conformational radius is equilibrium, with respect to σ_0, in order to transform it into the deformed state of the conformational ellipsoid with the semi-axes X_i, equilibrium with respect to σ_i. This determination means, that in Eq. (29) parameter σ_i should be replaced on σ_0, and values x_i should be replaced on X_i. Then we will obtain:

$$f_i = kT\frac{s(s-1)}{2}\left[-\frac{a^d(s/2)(2N)^2}{X_i\prod_i X_i} + \frac{X_i}{\sigma_0^2}\right] \qquad (66)$$

Substituting of $X_i = R_s \lambda_i$ in Eq. (66) we will obtain:

$$f_i = kT \frac{s(s-1)}{2} \frac{R_s}{\sigma_0^2} \frac{\lambda_i^2 \lambda_v - 1}{\lambda_i \lambda_v} \tag{67}$$

It follows from this, that in the accepted system of the mechanics signs the positive forces correspond to the stretching along i axe ($\lambda_i^2 \lambda_v > 1$), and the negative forces correspond to the compression ($\lambda_i^2 \lambda_v < 1$). That is why the main forces of deformation cannot be undefined, but they are ordered to the equation of the relationship (as it is following from the Eqs. (67) and (41)):

$$\sum_i f_i \lambda_i = 0 \tag{68}$$

A work of the deformation under the action of the all main forces le's describe by the expression:

$$A_{def} = \sum_i \int_{R_s}^{R_s \lambda_i} f_i dx_i \tag{69}$$

Using of the ratio (Eq. 67) and

$$d\lambda_v / \lambda_v = \sum_i d\lambda_i / \lambda_i \tag{70}$$

after the integration of Eq. (69) we will obtain the expression

$$A_{def} = kT \frac{d+2}{2} \left(\frac{R_s}{\sigma_0} \right)^2 \frac{s(s-1)}{2} \left(\frac{1}{\lambda_v} - 1 \right) \tag{71}$$

which is wholly identical to the obtained earlier (Eq. 45). This proves the correctness of the determination of the main forces accordingly to Eqs. (66) and (67).

2.2.5 DYNAMICAL PROPERTIES

2.2.5.1 CHARACTERISTIC TIME OF THE TRANSITION AND THE COEFFICIENT OF DIFFUSION

Characteristic time of the translational motion t_t^* of the strainless polymeric star (here we will be specialized on the analysis only of this situation) let's determine as a time needed for the transfer of its equilibrium frozen conformation on the characteristic distance R_s. At the transfer of sN of the links on a distance R_s it is necessary to do $(R_s / a)sN$ steps, every of which is realized for a time τ, which can be called as the characteristic time of the segmental motion.

Thereby,

$$t_t^* = (R_s / a)sN\tau \tag{72}$$

Substituting the Eq. (32) for R_s in Eq. (72) we will obtain

$$t_t^* = (2N)^{(d+5)/(d+2)}(s / 2)^{(d+3)/(d+2)}\tau \tag{73}$$

As it was shown in Ref. [32], the diffusion coefficient D at multivariate random transfer of the macromolecule is determined via expression:

$$D = \sigma_0^2 / 2t_t^* \tag{74}$$

which is the analog of the *Einstein's* equation $D = a^2 / 2\tau$ for low-molecular substances.

At the analysis of the directed transfer of macromolecule along i direction of the d-measured space, for example, under the action of gradient of chemical potential, it is necessary to use other determination of the diffusion coefficient:

$$D_i = \sigma_0^2 / 2dt_t^* \tag{75}$$

Substituting the Eqs. (73) and (31) in (75), finally we will obtain for the strainless polymeric star:

$$D_i = \frac{a^2}{2d\tau} \bigg/ (2N)^{3/(d+2)} (s/2)^{(d+3)/(d+2)} \tag{76}$$

2.2.5.2 CHARACTERISTIC TIME OF THE ROTATIONAL MOTION AND THE COEFFICIENT OF THE ELASTIC COMPONENT OF VISCOSITY

Viscous-elastic properties of the polymeric solutions suppose [32] the presence both of the frictional and elastic components of the measured effective viscosity. The elastic component of the viscosity is gradiently dependent value, depends also on a composition of the solution and on the coefficient of the elastic component of viscosity, which is determined via expression:

$$\eta_e^0 = \mu t_r^* \tag{77}$$

where: μ is the determined earlier shear modulus; t_r^* is the characteristic time of the rotational motion.

As the characteristic time of the rotational motion let's assume a time during which the strainless polymeric star into the frozen equilibrium conformational state will be rotated around any axis on the characteristic angle, equal to the one radian. Accordingly to this determination, the links, allocated from the rotation axis on a distance r, pass a way r for r/a steps and for time $(r/a)\,\tau$. Since the allocation of the all sN links in d-measured space is unknown, we use the following approach [33] for the estimation of t_r^*.

Let's design on the $(d-1)$-measured rotation plate the all sN links of the polymeric star. Conformational radius of the rotation plate is equal to R_s, but the numbers of the projections should accept such acceptable values N' and s', that to provide the value R_s in $(d-1)$-measured space. Then substituting the values $d-1$, N' and s' in Eq. (32) for R_s instead of d, N and s correspondingly, we will obtain

$$R_s = a(2N')^{3/(d+1)} (s'/2)^{1/(d+1)} \tag{78}$$

From the comparison of Eqs. (32) and (78) we will find the acceptable values of N' and s' in the rotation plate

$$2N' = (2N)^{(d+1)/(d+2)}, \quad s'/2 = (s/2)^{(d+1)/(d+2)} \tag{79}$$

Let's select on the rotation plate the linear polymeric chain by $2N'$ length, consisting of the pair of rays from the s'. Let n is the number of the link of presented given undefined chain from the rotation axis. Let's assume, that the distance r_n of this link from the rotation axis is ordered to the same distribution (Eq. 78), that is:

$$r_n = an^{3/(d+1)}(s'/2)^{1/(d+1)} \tag{80}$$

Although for the internal links of a chain this expression in not quite correct (*see* Chapter 1), but the following integration shows, that the main endowment into characteristic time of the rotation has the links with the numbers neared to $2N'$.

Under the plane of rotation turn on the one radian, the links with the numbers of n tract a way r_n for r_n/a steps and for time $(r_n/a)\tau$. For all $s'N'$ links, distributed upon $s'/2$ linear chains, the rotation time will be equal to $(s'/2)(r_n/a)\tau$. At the change of n on dn an increment of time consists of:

$$dt_r^* = (s'/2)(\tau/a)r_n dn \tag{81}$$

After the integration of Eq. (81) via the limits from $n = 1$ till $n = 2N'$ with taking into account of Eq. (78) we will obtain

$$t_r^* = ((d+1)/(d+4))(2N')^{(d+4)/(d+1)}(s'/2)^{(d+2)/(d+1)}\tau \tag{82}$$

By change of N' on N and s' on s accordingly to Eq. (79), we finally find

$$t_r^* = \frac{d+1}{d+4}(2N)^{(d+4)/(d+2)}\frac{s}{2}\tau \tag{83}$$

Next, using the Eqs. (65) and (50) for shear modulus of unstrained polymeric star $\lambda_v = 1$, we have

$$\mu = \chi(kT / a^d)(2N)^{4(1-d)/(d+2)}(s/2)^{4/(d+2)}(s-1) \tag{84}$$

Here: $\chi = (3 + \gamma(d-1))/(d-1)(1+\gamma)$

Combining the Eqs. (83) and (84) for determination of the coefficient of elastic component of viscosity in Eq. (77) we finally find accurate within a multiplier neared to the one:

$$\eta_e^0 \cong (kT / a^d)(2N)^{(8-3d)/(d+2)}(s/2)^{(d+6)/(d+2)}(s-1)\tau \tag{85}$$

Let's comment the obtained Eq. (76) for D_i and Eq. (85) for η_e^0, comparing them with the same expressions for D_i and η_e^0 of linear polymeric chains, consisting the same number of sN links. For this purpose let introduce the branching indexes upon Eq. (2) type in form $g_D = D_i / D_{iL}$ and $g_\eta = \eta_e^0 / \eta_{eL}^0$. For linear polymeric chains containing the sN links we have

$$D_{iL} = \frac{a^2}{2d\tau} \bigg/ (sN)^{3/(d+2)} \tag{86}$$

$$\eta_{eL}^0 \cong \frac{kT}{a^d}(sN)^{(8-3d)/(d+2)}\tau \tag{87}$$

Comparing the Eqs. (85) and (87) and also Eqs. (76) and (86) we have $g_D = (s/2)^{-d/(d+2)}$ and $g_\eta = (s/2)^{2(2d-1)/(d+2)}(s-1)$.

Thereby, at the same number of links, the polymeric stars are less mobile ($g_D < 1$), but have considerably more coefficient of the elastic component of viscosity $(g_\eta > 1)$, than the linear chains.

2.3   POLYMERIC STARS IN CONCENTRATED SOLUTION

2.3.1   SELF-AVOIDING RANDOM WALKS STATISTICS

In concentrated solution the conformational volumes of polymeric stars are overlapped in accordance with the condition $\rho > \rho^*$, in which ρ is a density of the solution upon polymer, ρ^* is the critical density corresponding to the start of the conformational volumes overlapping. Due to the polymeric star cannot be considered as an independent subsystem, it is necessary to consider the all set of the polymeric stars in conformational volume of the separated system.

Let's introduce the screen cubic d-measured space containing of m intertwining between themselves uniform polymeric stars. Let's separate from them any undefined star, and in it – any undefined pair of rays, forming the linear chain by $2N$ length. Its phantom *Gaussian* walks are ordered to the same distribution law (Eq. 13) with the same normalization condition (Eq. 10).

However, the probability $\omega(z)$ of the msN differed links distribution upon Z identical cells at the condition, that in every cell there is not more than one link, will be determined by a new expression:

$$\omega(z) = \frac{Z!}{(Z-msN)!}(\frac{1}{Z})^{msN}(1-\frac{1}{Z})^{Z-msN}$$

(88)

It follows from this, that the probability of the self-avoiding walks for undefined chosen pair of rays will be equal:

$$\omega(2N) = \left(\omega(Z)\right)^{2/ms} \omega(n)$$

(89)

Here the index $2/ms$ was obtained as the ratio $2N/msN$. Combining the Eqs. (13) and (88) into (89) and using as same principles as at the derivation of the Eq. (16), we will obtain into the asymptotic limit from Eq. (89):

$$\omega(2N) = \exp\left\{-\frac{(ms/2)(2N)^2}{Z} - \frac{1}{2}\sum_i \frac{s_i^2}{n_i}\right\}$$

(90)

The transition to the d-measured space is realizable with the use of the same previous Eqs. (17)–(20), accordingly to which

$$\omega(2N) = \exp\left\{-\frac{a^d(ms/2)(2N)^2}{\prod_i x_i} - \frac{1}{2}\sum_i \frac{x_i^2}{\sigma_i^2}\right\}$$

(91)

This distribution is true for any undefined pair of rays of any undefined star in general space $z = \prod_i x_i/a^d$.

2.3.2 THERMODYNAMICS OF CONFORMATION AND DEFORMATION OF THE INTERTWINING POLYMERIC STARS

The numbers of configurations for pair of rays, polymeric star and m intertwined between themselves in m-ball polymeric stars let's determine, correspondingly, by the expressions:

$$L(2N) = L_{max}(2N)\omega(2N) \tag{92}$$

$$L(sN) = \left(L(2N)\right)^{s(s-1)/2} \tag{93}$$

$$L(msN) = \left(L(2N)\right)^{ms(s-1)/2} \tag{94}$$

Therefore, in accordance with *Boltzmann*, general entropy of the self-avoiding walks of the intertwined polymeric stars in m-ball will be equal:

$$S_0 = k\left(ms(s-1)/2\right)\ln L_{max}(2N) + k\left(ms(s-1)/2\right)\ln\omega(2N) \tag{95}$$

As same as earlier, only the second term in Eq. (95) we accept as the entropy of the conformation of m-ball of polymeric stars or as entropy of their self-organization:

$$S = k\left(ms(s-1)/2\right)\ln\omega(2N) \tag{96}$$

At the absence of the energetic effects, a free energy of the conformation will be equal to:

$$F_m(x) = kT\frac{ms(s-1)}{2}\exp\left\{\frac{a^d(ms/2)(2N)^2}{\prod_i x_i} + \frac{1}{2}\sum_i \frac{x_i^2}{\sigma_i^2}\right\} \tag{97}$$

From the all possible states of the polymeric stars in m-ball let's choice the most probable or thermodynamically equilibrium states in accordance with the condition $\partial F_m(x)/\partial x_i = 0$. Differentiating of Eq. (98), we will obtain:

$$\partial F_m(x)/\partial x_i = kT\frac{ms(s-1)}{2}\exp\left\{-\frac{a^d(ms/2)(2N)^2}{x_i\prod_i x_i}+\frac{x_i}{\sigma_i^2}\right\} \tag{98}$$

Equaling the right parts of Eq. (98) to zero for all $i = 1,d$ and solving the obtained system of the algebraic equations, let's find the equilibrium semi-axes of the conformational ellipsoid, general for any pair of rays, any polymeric star and m-ball in the large:

$$X_i = \sigma_i\left[\frac{a^d(ms/2)(2N)^2}{\prod_i \sigma_i}\right]^{1/(d+2)} \tag{99}$$

At the equiprobability of walks upon the all directions of d-measured space that is reflected by the condition of $\sigma_i = \sigma_0$ accordingly to Eq. (31), we have the spherical conformational volume with the radius:

$$R_m = a(2N)^{3/(d+2)}(ms/2)^{1/(d+2)} \tag{100}$$

Deformation of the m-ball at its transition from the spherical (unstrained) conformational state into the ellipsoid with the semi-axes X_i let's express via the multiplicities of the linear λ_i and volumetric λ_v deformation via ratios:

$$\lambda_i = X_i/R_m, \qquad \lambda_v = \prod_i X_i\Big/R_m^d = \prod_i \lambda_i \tag{101}$$

which are also ordered to the relationship of Eq. (41).

Next, substituting the equilibrium values $X_i = R_m\lambda_i$ in Eq. (97) we will find the equilibrium free energy of the strained m-ball:

$$F_m(\lambda) = kT\frac{d+2}{2}\left(\frac{R_m}{\sigma_0}\right)^2\frac{ms(s-1)}{2}/\lambda_v \tag{102}$$

It follows from this, that in the system of the mechanics signs a work of the transition of the unstrained m-ball with $\lambda_v = 1$ into the deformated state of the conformational ellipsoid with $\lambda_v < 1$ will be equal to:

$$\Delta F_{def} = kT \frac{d+2}{2} \left(\frac{R_m}{\sigma_0} \right)^2 \frac{ms(s-1)}{2} \left(\frac{1}{\lambda_v} - 1 \right) \tag{103}$$

Next, determining the conformation pressure via the same thermodynamic ratio (Eq. 47), in which the conformational volume of m-ball in general case is equal to $V = R_m^d \lambda_v$, we find

$$P = kT \frac{d+2}{2} \left(\frac{R_m}{\sigma_0} \right)^2 \frac{ms(s-1)}{2} \Big/ R_m^d \lambda_v^2 \tag{104}$$

Again we have $P = F/V$ and the equation of the conformational state of m-ball for intertwined polymeric stars:

$$PV^2 = FV = const, \tag{105}$$

$$const = kT \frac{d+2}{2} \left(\frac{R_m}{\sigma_0} \right)^2 \frac{ms(s-1)}{2} R_m^d \tag{106}$$

2.3.3 MODULUS OF ELASTICITY OF M-BALL FOR INTERTWINING POLYMERIC STARS

An equation of the deformation for the m-ball of the intertwined polymeric stars we write in the same general differential form (Eq. 53) and transform it relatively to the *Young's* modulus into the form (Eq. 54). Corresponding derivatives we find via the differencing of (98) with the following substitution of the equilibrium values $x_i = X_i = R_m \lambda_i$:

$$\partial f_i / \partial x_i = 3kTms(s-1)/2\sigma_0^2 \lambda_i^2 \lambda_v, \tag{107}$$

$$\partial f_i / \partial x_j = \partial f_j / \partial x_i = kTms(s-1)/2\sigma_0^2 \lambda_i \lambda_j \lambda_v \tag{108}$$

Substituting of these expressions in Eq. (54), we will obtain:

$$Y = kT(3 + \gamma(d-1))\left(\frac{R_m}{\sigma_0}\right)^2 \frac{ms(s-1)}{2} \bigg/ R_m^d \lambda_v^2 \tag{109}$$

Comparing of Eq. (109) with the Eq. (104) for P, we have the relationship between Y and P:

$$Y = 2(3 + \gamma(d-1))P / (d+2) \tag{110}$$

From the other hand, determining of Y via the volumetric modulus E = 2P by the ratio (Eq. 60), we will again obtain the relationship by Eq. (62) type, comparing of which with the Eq. (110), we will obtain the expression for the *Poisson* coefficient in the well-known form Eq. (63). So, the *Poisson* coefficient both for the linear chains and for the polymeric stars in diluted and concentrated solutions is the universal function only on the *Euclidian* space.

Using the Eq. (64), let's express the shear modulus μ of the m-ball of the intertwined polymeric stars via the conformation pressure in the known form:

$$\mu = \frac{2(3 + \gamma(d-1))}{(d+2)(d-1)(1+\gamma)} P \tag{111}$$

2.3.4 MAIN FORCES AND A WORK OF THE DEFORMATION OF M-BALL OF THE INTERTWINING POLYMERIC STAR

The main forces of the deformation let's again determine as the forces, which should be applied to the m-ball, in order to transfer it from the unstrained state equilibrated with respect to the σ_0, into the deformated state, equilibrated with respect to the σ_i. That is why it is necessary again to substitute instead of σ_i the value σ_0 in the expression $f_i = \partial F_m / \partial x_i$ accordingly to (98), and x_i to change on X_i, that is

$$f_i = kTm \frac{s(s-1)}{2} \left[-\frac{a^d m(s/2)(2N)^2}{x_i \prod_i x_i} + \frac{x_i}{\sigma_0^2} \right] \tag{112}$$

After the substitution $X_i = R_m \lambda_i$, we have

$$f_i = kT \frac{ms(s-1)}{2} \frac{R_m}{\sigma_0^2} \frac{\lambda_i^2 \lambda_v - 1}{\lambda_i \lambda_v} \tag{113}$$

Describing the deformation work of the m-ball by the same equation of mechanics (Eq. 69) with the use of the Eq. (113) for the main forces, we will obtain again the expression $A_{def} = \Delta F_{def}$ in accordance with the Eq. (103).

2.3.5 DETERMINATION OF m, R_m, F_m AND P AS THE EXPLICIT FUNCTIONS S, N AND CONCENTRATION ρ OF POLYMERIC STARS IN SOLUTION

Concentration (density of solution upon polymer) of polymeric stars in concentrated solution is equal to

$$\rho = M_0 ms N / N_A R_m^d \tag{114}$$

where M_0 is the molar mass of a link. At $m = 1$ we have the critical concentration, to which the beginning of the polymeric stars conformational volumes overlapping corresponds:

$$\rho^* = M_0 s N / N_A R_s^d \tag{115}$$

From the comparison of Eqs. (114) and (115) follows

$$m^{2/(d+2)} = \rho / \rho^* \tag{116}$$

Determining the density in the volume of a link via expression

$$\rho_0 = M_0 / N_A a^d \tag{117}$$

we find

$$\rho^* = \rho_0 (2N)^{2(1-d)/(d+2)} (s/2)^{2/(d+2)} \tag{118}$$

This permits to write for m and R_m the expressions

$$m = (2N)^{d-1} (\rho / \rho_0)^{(d+2)/2} (s/2)^{-1},$$ (119)

$$R_m = a(2N)(\rho / \rho_0)^{1/2}.$$ (120)

via substituting of which in Eqs. (102) and (104), we will find

$$F_m(\lambda) = ((d+2)/2)kT(2N)^d (\rho / \rho_0)^{(d+4)/2} (s-1)/\lambda_v$$ (121)

$$P = \frac{d+2}{2} \frac{kT}{a^d} \left(\frac{\rho}{\rho_0} \right)^2 (s-1)/\lambda_v^2$$ (122)

We can easy find the explicit functions of the modulus of elasticity on N, s and ρ via P in accordance with (122).

2.3.6 A WORK OF THE INTERTWINING OF POLYMERIC STARS

A change of free conformation energy at the transition of the m polymeric stars from the diluted solution into the concentrated one is equal to $F_m - mF_s$. In calculation per one polymeric chain we have:

$$\Delta F_p = (F_m - mF_s)/m$$ (123)

In a system of the mechanics signs this value determines a work of the polymeric stars intertwining as a work of the polymeric star transfer from the diluted solution in concentrated one. After the substitution in Eq. (123) of the expressions Eqs. (42), and (102) we will obtain:

$$\Delta F_p = \frac{d+2}{2} kT \frac{s(s-1)}{2} \left(\frac{R_s}{\sigma_0} \right)^2 \left(m^{2/(d+2)} - 1 \right)$$ (124)

where $m^{2(d+2)} = \rho/\rho^*$. From the other hand, expressing the conformational volumes of polymeric chains in diluted solution $V_s = mR_s^d$, and in concentrated $V_m = R_m^d$, we have

$$\lambda_v = V_m / V_s = m^{-2/(d+2)}$$ (125)

Therefore, a work of the polymeric stars overlapping into the m-ball represents by itself a work of the conformational volume compression at the transfer of the star from the diluted solution into the concentrated one:

$$\Delta F_p = \frac{d+2}{2} kT \frac{s(s-1)}{2} \left(\frac{R_s}{\sigma_0}\right)^2 (1/\lambda_v - 1) \tag{126}$$

2.3.7 DYNAMICAL PROPERTIES OF POLYMERIC STARS IN CONCENTRATED SOLUTION

2.3.7.1 CHARACTERISTIC TIME OF THE TRANSITION AND A DIFFUSION COEFFICIENT

Characteristic time of the translational motion of polymeric stars in concentrated solution let's determine as a time t_{mt}^*, for which the m-ball of the intertwined polymeric stars with the frozen equilibrium conformation will be displaced on the effective distance R_m consequently of the random walks upon the all d directions of space. The next expression corresponds to this determination:

$$t_{mt}^* = (R_m / a) msN\tau \tag{127}$$

in which as before τ is the characteristic time of the segmental motion. Substituting in this expression the Eqs. (119) and (120), determining the m and R_m, we will obtain:

$$t_{mt}^* = (2N)^{d+1} (\rho/\rho_0)^{(d+3)/2} \tau \tag{128}$$

A diffusion coefficient for the chosen direction, determined earlier by the Eq. (75), will be described via expression:

$$D_i = \frac{a^2}{2d\tau} (2N)^{-d} (\rho/\rho_0)^{-(d+3)/2} \tag{129}$$

2.3.7.2 CHARACTERISTIC TIME OF THE ROTATIONAL MOTION AND THE COEFFICIENT OF THE ELASTIC COMPONENT OF VISCOSITY

Characteristic time t^*_{mr} of the rotational motion of the m-ball of the intertwined polymeric stars let's determine as a time, needed for the turn of the frozen equilibrium conformation of m-ball on the elementary angle equal to the one radian.

Let's select the rotation plate by dimensionality d–1 with the same conformational radius R_m and blueprint on it the all msN links of the m-ball. Obtained projections N', s' and m' are ordered to the $SARW$ statistics, that is why the conformation radius in d–1 plate can be write as follow:

$$R_m = \left(2N'\right)^{3/(d+1)} ((ms/2)')^{1/(d+1)} \tag{130}$$

Comparing the Eq. (130) and (100), we determine the relationships

$$(2N)' = (2N)^{(d+1)/(d+2)}, \quad (ms)' = (ms)^{(d+1)/(d+2)} \tag{131}$$

Let's select a chain by the $2N'$ length from the general numbers of the projections. Assuming that the internal links of this chain with the numbers of n from the rotation center are ordered to the same regularity (Eq. 130), their rotation radius will be expressed via ratio:

$$r_n = an^{3/(d+1)}\left((ms)'/2\right)^{1/(d+1)} \tag{132}$$

At the turn on an angle by the one radian these projections with the numbers n from the rotation center pass the distance r_n for r_n/a steps and for time $(r_n/a)\tau$. At the change of n on dn an increment of time consists of $dt^*_{1r} = (r_n/a)\tau dn$. By integrating of this expression from $n = 1$ till $n = 2N'$, we will obtain:

$$t^*_{1r} = \frac{d+1}{d+4}\left(2N'\right)^{(d+4)/(d+1)}\left(\frac{(ms)'}{2}\right)^{1/(d+1)}\tau \tag{133}$$

For the all $(ms)^{1/2}$ chains we have $t_{mr}^* = t_{1r}^*(ms)^{y}/2$. Taking into account of the Eq. (131) we finally find:

$$t_{mr}^* = \frac{d+1}{d+4}(2N)^{d+2/(d+2)}\left(\frac{\rho}{\rho_0}\right)^{(d+2)/2}\tau \qquad (134)$$

We determine the coefficient of the elastic component of viscosity of concentrated solution of polymeric stars $\eta_{me}^0 = \mu t_{mr}^*$ via the characteristic time t_{mr}^* and the shear modulus μ. Using the Eqs. (111) and (122) at λ_v =1 for μ and Eq. (134) for t_{mr}^*, we find:

$$\eta_{me}^0 = \chi \frac{kT}{a^d}(2N)^{d+2/(d+2)}\left(\frac{\rho}{\rho_0}\right)^{(d+6)/2}(s-1)\tau \qquad (135)$$

where: the value $\chi = (d+1)(3+\gamma(d-1))/(d+4)(d-1)(1+\gamma)$ for $d=3$ space neared to the one, that is why it can be written:

$$\eta_{me}^0 \cong \frac{kT}{a^3}(2N)^{3,4}\left(\frac{\rho}{\rho_0}\right)^{4,5}(s-1)\tau \qquad (136)$$

2.4 CONCLUSION

Self-avoiding random walks statistics completely describes the thermodynamic and dynamic properties of the polymeric stars in diluted solutions as the function on a length and the number of rays; in concentrated solutions additionally as the function on the concentration of polymer.

KEYWORDS

- **diluted and concentrated solutions**
- **polymeric stars**
- **self-avoiding random walks statistics**
- **viscosity**

REFERENCES

1. Nayak S., Lyon L. A., *Angew. Chem. Int. Ed.* 2005, v. 44. № 9, p. 7686.
2. Hammond P. T., *Adv. Mater.*, 2004, v. 16, № 15, p. 1271.
3. Stansel–Rosenbaum M., Davis T. P., *Angew. Chemun.*, 2001, v. 113, № 12, p.35.
4. Bauer B. J., Fetters I. J., Graessly W. W., Hadjichristidis N., Quack G.F., *Macromolecules*, 1989, v. 22, № 5, p. 2337.
5. Hutchings L. R., Richards R. W., *Macromolecules*, 1999, v. 32, № 3, p. 880.
6. Hutchings L. R., Richards R. W., Reynolds S. W., Thompson R. L., *Macromolecules*, 2001, v. 34, № 16, p. 5571.
7. Phyllipov A. P., Romanova O. A., Vynogradova L. V., *High Molecular Compounds A.*, 2010, v. 52, № 3, p. 371 *(in Russian)*.
8. Daoud M., Cotton J. P., *J. Phys. (France)*, 1982, v. 43, № 3, p.531.
9. Birshtein T. M., Zhulina E. B., *Polymer*, 1984, v. 25, p. 1453.
10. Alexander S., *J. Phys. (France)*, 1977, v. 38, № 8, p. 983.
11. de Gennes P. G., *Macromolecules*, 1980, v. 13, № 5, p. 1069.
12. Birshtain T. M., Merkuryeva A. A., Leermakers F. A., Rud' O. V., *High Molecular Compounds A.*, 2008, v. 50, № 9, p. 1673.
13. Duplantier B., *Stat. Phys.*, 1989, v. 54, p.581.
14. Guida R., Zinn-Justin J., *J. Phys. A,*, 1998, v. 31, p. 8104.
15. von Ferber Ch., Holovatch Yu., *Condens. Matter Phys.*, 2002, v. 5, № 1, p. 117.
16. Miyake A., Freed K. F., *Macromolecules*, 1983, v. 16, № 7, p. 1228.
17. Miyake A., Freed K. F., *Macromolecules*, 1984, v. 17, № 4, p. 678
18. Schafer L., von Ferber C., Lehr U., Duplantier B., *Nucl. Phys. B,*, 1992, v. 374, № 3, p. 473.
19. Murat M., Crest G. S., *Macromolecules*, 1989, v. 22, № 10, p. 4054.
20. Crest G. S., Kremer K., Witten T. A., *Macromolecules*, 1987, v. 20, № 6, p. 1376.
21. Berret A. J., Tremain D. L., *Macromolecules*, 1987, v. 20, № 7, p.1687.
22. Batoulis J., Kremer K., *Macromolecules*, 1989, v. 22, № 11, p. 4277.
23. Ohno K., *Condens. Matter Phys.*, 2002, v. 5, № 1(29), p. 15.
24. des Cloizeaux J., *Phys. Rev.*, 1974, A 10, p. 1565.
25. de Gennes P. G. Scaling Concepts in Polymer Physics, *Cornell Universiti Pres., Ithaka*, 1979.
26. des Cloizeaux J., *J. Physique*, 1980, v. 41, p. 223.
27. Redner S., *J. Phys. A: Math. Gen.*, 1980, v. 13, p. 3525.
28. Ohno K., Binder K., *J. Chem. Phys.*, 1991, v. 95, № 7, p. 5459.
29. Medvedevskikh Yu. G., *Condens. Matter Phys.*, 2001, v. 4, № 2, p. p. 209, 219.
30. Medvedevskikh Yu. G., *J. Appl. Polym. Sci.*, 2008, v. 109, p. 2472.
31. Hans-Georg E. An Introduction to Polymer Science,, *WCH, Weinheim*, 1997.
32. Medvedevskikh Yu. G. In: Conformation of Macromolecules. Thermodynamic and Kinetic Demonstrations / *Medvedevskikh Yu. G. Voronov S. A., Zaikov G. E. (Ed.), Nova Science Publishers, Inc. New York*, 2007, p. 107.
33. Medvedevskikh Yu. G. In: Conformation of Macromolecules. Thermodynamic and Kinetic Demonstrations / *Medvedevskikh Yu. G. Voronov S. A., Zaikov G. E. (Ed.), Nova Science Publishers, Inc. New York*, 2007, p. 125.

CHAPTER 3

THE NUMBER OF CONFIGURATIONS OF POLYMERIC CHAIN IN THE SELF-AVOIDING RANDOMWALKS STATISTICS

YU. G. MEDVEDEVSKIKH

CONTENTS

SUMMARY

The number of configurations L of the linear polymeric chain accurate within the constant multiplier neared to unit is unambiguously determined via the average variance z of the step of SARW trajectory: $L \approx z^N$. Probabilistic analysis of the SARW trajectories leads to the expression $z = (2d - 1)(1 - p)$, in which p is the average upon the all SARW trajectories probability to discover the neighboring cell by occupied. The SARW statistics leads to the ratio $z = (2d - 1)exp\left\{-\dfrac{d+2}{2}\theta\right\}$, in which θ is an average occupancy cell upon the conformational volume. From the comparison of these expressions the next relationship follows: $p = 1 - exp\left\{-\dfrac{d+2}{2}\theta\right\}$. The three last expressions are retained for the linear chains and polymeric stars into diluted and concentrated solutions, ideal and real ones. The number of configurations L_{2N} for any pair of rays of the polymeric star with the s rays by the N length is determined by the expression $L_{2N} = z^{2N}$, and for the whole star $L_{sN} = z^{s(s-1)N}$.

3.1   INTRODUCTION

The number of configurations L of a polymeric chain is one among methods of its conformational state realization. Under this sense L is the statistical analog of the important thermodynamical characteristic of the conformational state of a polymeric chain, *namely* its entropy S: $S = klnL$, where k is the *Boltzmann's* constant.

The first results of the numerical estimation L for linear polymeric chain at little values of number of its inks N with the use of the *Monte-Carlo* method were interpreted in a form of the scaling dependence [1, 2]:

$$L \approx z^N N^{\gamma-1} \qquad (1)$$

Parameter z was determined as un-universal constant *or* effective coordinating number of d-measured cubic lattice, in space of which the trajectory of self-avoiding random walks *(SARW)* of the polymeric chain is constructed; γ is the universal scaling index, depending only on the dimension d of the screen space.

The first estimations of values $z = 4.68$ and $\gamma = 1.16$ at $d = 3$ later were made more exact: $z = 4.6853$ [3], $\gamma = 1.1596$ [4].

For polymeric star consisting of s rays by equal length N, the number of configurations is also postulated by the scaling expression of type (1) [5, 6]:

$$L \approx z^{sN} N^{\gamma_s - 1} \tag{2}$$

With the use of the calculations performed by the methods of group renormalization of field theory [7] and by the *Monte-Carlo* method [8–10] it was shown, that the scaling index γ_s of the polymeric star very nontrivially depends on the number of the rays: under the s increasing the index γ_s firstly slowly is decreased to zero (at $s \sim 7$), and after that under $s > 7$ it's sharply decreased taking the negative values up to $\gamma_s = -29$ at $s = 32$ [8]. Such values are badly agreed with the physical interpretation of the scaling index. Probably, this caused by the absence of numerical estimations of z parameter and its possible dependence on s and N.

In connection with this fact let us note, that the both Eqs. (1) and (2) represent the number of the configurations of polymeric chain as two cofactors, absolutely different upon its "weight". Let us estimate of their weights accordingly to the Eq. (1) for linear polymeric chain using the presented above values $z = 4.68$ and $\gamma = 1.16$ for the reference point. At $N = 50$ we will obtain: $L = 4.68^{50}\ 50^{0.16} = (3.3\ 10^{33})(1.9)$. So, the main factor determining the value L, is the co-factor z^N, against the background of which the co-factor $N^{\gamma - 1}$ has an insignificant role. This is visualized also under the comparison of their endowment into the entropy of conformation which is proportional to $N \ln z = 77.1$ and $(\gamma - 1) \ln N = 0.6$ correspondingly. As we can see, these endowments are differed on two orders; under the N increasing the difference will be just only increased.

That is why in the presented paper the all attention will be paid into the analysis of z parameter of linear chains and polymeric stars into diluted and concentrated, ideal and real solutions.

3.2   AN AVERAGE VARIANCE OF TRAJECTORIES STEP OF SARW AND THEIR NUMBER

Any random configuration of the polymeric chain can be considered as the trajectory of *SARW* in N steps into the d-measured screen space with the size of the cubic cell, which is equal to the length of the monomeric link of a chain. The connectedness of the monomeric links into a polymeric chain makes the first and very important contingency on the trajectory of the *SARW*, *namely* the prohibition of step backwards [8]. That is why only the first step has the $2d$ methods or variants of transition into the neighboring cells; the second and the following steps can to have not more than $2d-1$ variants of the transition. If among $2d-1$ of the neighboring cells the n are occupied, then the number of the variants of transition on presented step is equal to the number of unoccupied or empty cells, that is $2d-1-n$. The number n can be changed via the limits from 0 to $2d-1$. The last means that the trajectory of *SARW* finds oneself into the trap with the absence of variants of the transition into the neighboring cells. This case is very interesting for the kinetics of the macroradical propagation at the polymerization, since represents by itself the monomolecular chain termination [*11*]. Under analysis of the number of configurations of polymeric chain the value n can be limited by a number of $2d-2$ which makes the following step by monovariant, and therefore, by possible.

Let introduce the average probability p_n of that the n of the neighboring cells occupied. Then the average variance of step z_p for the all trajectories of *SARW* will be equal:

$$z_p = \sum_{n=0}^{2d-2}(2d-1-n)p_n \tag{3}$$

Every step of the *SARW* trajectory represents by itself the $2d-1$ independent tests on occupancy of the neighboring cells and random transition into the one among free cells. Therefore, in accordance with the theorem about the repeated tests the average probability of that among of $2d-1$ of the neighboring cells exactly n will be occupied, and $2d-1-n$ will be vacant, is ordered to the binomial distribution law:

$$p_n = C_{2d-1}^n p^n (1-p)^{2d-1-n}$$ (4)

Here the binomial coefficients are described by the expressions:

$$C_{2d-1}^n = (2d-1) / n!(2d-1-n) ,$$ (5)

p is the mathematical expectation or the average upon the all *SARW* trajectories probability of the occupancy of the one cell.

Combining the Eqs. (3) and (4), we will obtain:

$$z_p = \sum_{n=0}^{2d-2} (2d-1-n) C_{2d-1}^n p^n (1-p)^{2d-1-n}$$ (6)

Due to the probabilistic or stochastic character of the *SARW* trajectories the Eq. (6) is true only at $d \geq 2$. For the one-dimensional space only the first step has the variance $2d$, the rest of $N-1$ steps determined, in other words are not stochastic, and that is why cannot be described by the Eq. (6).

Since in accordance with the determination of Eq. (6) z_p is the average variance of the step of trajectories in $N-1$ steps, and the first step has the $2d$ variants, a general number L_p of different trajectories or configurations of polymeric chain upon the property of the multiplicatively will be equal to:

$$L_p = \frac{2d}{z_p} z_p^N$$ (7)

The Eq. (6) permits to analyze the endowments of steps with the variance $2d-1-n$ into the average variance of step z_p of *SARW* trajectories. Let us illustrate of this fact on the example of $d = 3$ space, for which the Eq. (6) takes the form:

$$z_p = 5(1-p)^5 + 20(1-p)^4 + 30p^2(1-p)^3 + 20p^3(1-p)^2 + 5p^4(1-p)$$ (8)

Under two random values $p_1 = 0.1$ and $p_2 = 0.01$ we have correspondingly:

$$z_{p1} = 2,9525 + 1,3122 + 0,2187 + 0,0162 + 0,0004 \cong 4,50$$

$$z_{p2} = 4,7549 + 01921 + 0,0029 + 2 \bullet 10^{-5} + 5 \bullet 10^{-7} \cong 4,95$$

Here the first terms give the endowment into z_p steps with $n = 0$, the second ones – with $n = 1$, etc. As we can see, under the p decreasing the average variance of a step is increased at the expense of the sharp steps endowment decreasing with $n \geq 1$ and sharp increasing of the steps endowment with $n = 0$.

However, if don't use of this detail information, but to be concentrated only on the value z_p, it can be find without taking into account of the binomial distribution law. Really, since the p is an average upon the all trajectories probability to discover the occupied cell, the mathematical expectation of the number of occupied cells at $2d-1$ independent tests will be equal to $(2d-1)p$. Correspondingly, the mathematical expectation of the number of empty cells under the same $2d-1$ independent tests will be equal to $(2d-1)(1-p)$.

Exactly this number determines the average variance of a step of the *SARW* trajectories:

$$z_p = (2d - 1)(1 - p) \tag{9}$$

By substituting in this expression the previous undefined values $p_1 = 0.1$ and $p_2 = 0.01$, we will again obtain $z_{p1} = 4.5$ and $z_{p2} = 4.95$.

At $p < 1$ the Eq. (9) can be written in the form

$$z_p = (2d - 1)\exp(-p) \tag{10}$$

As we can see from the Eqs. (9) and (10), at $p = 0$ the average variance of a step of the *SARW* trajectories takes its maximal value: $z_p = 2d-1$. Correspondingly, the maximal number of the *SARW* trajectories or the configurations of polymeric chain replies to a case $p = 0$:

$$L_{max} = \frac{2d}{2d - 1}(2d - 1)^N \tag{11}$$

Condition $p = 0$ points on the single contingencies, superposed on the *SARW* trajectories: any among their steps cannot be returned due to the connectedness of the monomeric links into the chain. The remaining contingencies of the self-avoiding random walks lead to the condition $p > 0$.

Performed analysis shows, that the average variance of a step of the *SARW* trajectories is the universal function only on two parameters, *namely d* and *p*. However, into presented approach the parameter *p* is not determined. Evidently, it should be depending on the type of a polymeric chain (e.g., linear or star-like), the length of a chain or the rays and their number, the concentration of a polymer into solution and its thermodynamical properties (ideal or real). An analysis of the influence of these factors on parameter *p* let's carried out within the strict *SARW* statistics [*12, 13*], which considers the conformation of a polymeric chain as the result of the statistical average upon the all possible configurations with taking into account the probability of their realization.

3.3   AN AVARAGE VARIANCE OF THE STEP IN THE SARW STATISTICS

3.3.1   *LINEAR POLYMERIC CHAINS*

3.3.1.1   *DILUTED SOLUTIONS, IDEAL AND REAL ONES*

The *SARW* statistics of linear polymeric chain into diluted solution determines [*12*] the density of distribution $\omega(\lambda)$, to which corresponds the probability $\omega(\lambda)\Pi_i d\lambda_i$ $i = 1, d$ of that the *SARW* trajectory by its last step hits into the volume of the elementary layer $R_f^d \Pi_i d\lambda_i$ on the surface of the equilibrium conformational ellipsoid with the semi-axes's $X_i = R_f \lambda_i$, in a form

$$\omega(\lambda) = \exp\left\{ -\left(\frac{R_f}{\sigma_0} \right)^2 \left(\frac{1}{\Pi_i \lambda_i} + \frac{1}{2} \sum_i \lambda_i^2 \right) \right\} \tag{12}$$

where: $\sigma_0^2 = a^2 N$ is the root-mean-square deviation of the *Gaussian* part (Eq. 12); R_f is the most probable radius of the polymeric chain conformation into the ideal diluted solution *or* the radius of the un-deformed *Flory* ball:

$$R_f = aN^{3/(d+2)}$$

(13)

It follows from this

$$\left(R_f / \sigma_0\right)^2 = N^{(4-d)(d+2)}$$

(14)

Parameters λ_i are the multiplication factors of a linear deformation of the *Flory* ball along the corresponding axis's of d-measured space; $\Pi_i \lambda_i = \lambda_v$ is the multiplication factor of the volumetric deformation. For a polymeric chain into the ideal solution the all $\lambda_i = 1$ and $\lambda_v = 1$. Under any deformations of the *Flory* ball its conformational volume is decreased, that is why in the real solution $\lambda_v < 1$.

Parameters λ_i cannot take the unconditioned values, since they are connected via the ratio

$$\sum_i \lambda_i^2 = d / \Pi_i \lambda_i$$

(15)

This permits to write the Eq. (12) in more convenient form for the following analysis:

$$\omega(\lambda) = \exp\left\{-\frac{d+2}{2}\left(\frac{R_f}{\sigma_0}\right)^2 / \lambda_v\right\}$$

(16)

Since the density of distribution represents the result of the statistical average upon the all possible configurations of a polymeric chain with taking into account of the probability of their realization, it can be considered as the ratio of number L_ω of the *SARW* trajectories, realizing the presented conformational state, to the maximally possible number of the trajectories which limited only by the connectedness of the links into a chain:

$$\omega(\lambda) = L_\omega / L_{max}$$

(17)

Taking into account the Eq. (11), it follows from this:

$$L_\omega = \frac{2d}{2d-1}(2d-1)^N \omega(\lambda) \tag{18}$$

By substituting of the Eqs. (14) and (16) into Eq. (18), we will obtain

$$L_\omega = \frac{2d}{2d-1}\left[(2d-1)\exp\left\{-\frac{d+2}{2}N^{2(1-d)(d+2)}/\lambda_v\right\}\right]^N. \tag{19}$$

This permits to write

$$L_\omega = \frac{2d}{2d-1}z_\omega^N, \tag{20}$$

where z_ω is an average variance of a step in the $SARW$ statistics:

$$z_\omega = (2d-1)\exp\left\{-\frac{d+2}{2}N^{2(1-d)(d+2)}/\lambda_v\right\} \tag{21}$$

Next let's introduce an average occupancy of a cell into the conformational volume of a polymeric chain via the ratio

$$\theta = \frac{a^d N}{R_f^d \lambda_v} \tag{22}$$

from which with taking into account of Eq. (13) follows

$$\theta = N^{2(1-d)/(d+2)}/\lambda_v \tag{23}$$

Comparing the Eqs. (21) and (23), we find

$$z_\omega = (2d-1)\exp\left\{-\frac{d+2}{2}\theta\right\} \tag{24}$$

Definitionally on Eq. (22) θ is the probability to discover the cell occupied into conformational volume of linear polymeric chain, and under this sense it could be equated to p. However, such assumption doesn't take into account, that the Eq. (22) into the evident form supposes the uniform distribution of the links of a chain into its conformational volume. Any among *SARW* trajectories cannot be uniformly distributed upon the whole conformational volume and that is why due to the local character of the *SARW* trajectories the condition $p > \theta$ should be performed. Comparing the Eqs. (9) and (24) and taking into account that the both of them should represent the same physical value $z_p = z_\omega = z$, in general case we obtain the following relationship:

$$1 - p = \exp\left\{-\frac{d+2}{2}\theta\right\} \qquad (25)$$

the partial case of which under $p < 1$ and $\theta < 1$ is the ratio

$$p = \frac{d+2}{2}\theta \qquad (26)$$

Let's note that although the *SARW* statistics is based on the indispensable condition $N > 1$, both conditions $p < 1$ and $\theta < 1$ can simultaneously and exactly don't perform. That is why more general Eq. (25) will be used into the following calculations.

In accordance with the Eqs. (7) and (20) under condition $z_p = z_\omega = z$ between L_p and L_ω the difference in co-factors $2d/z_p$ and $2d/(2d-1)$ neared to 1 and having a little significance at the co-factor z^N is kept. That is why without a great error it can be taken that $L_p = L_\omega = L$, and L can be expressed via the ratio

$$L \cong z^N \qquad (27)$$

and finally the average variance of a step of trajectories of linear polymerization for a chain in the *SARW* statistics can be determined via expression

$$z = (2d-1)\exp\left\{-\frac{d+2}{2}\theta\right\}. \qquad (28)$$

For illustration of the dependence of θ, p and z on N, d and λ_v in Table 1 there are their calculated values upon the Eqs. (23), (25) and (28).

TABLE 1   Calculated values of an average part of the configurational volume on conformational one under different variants.

N	$d = 2, \lambda_v = 1$			$d = 3, \lambda_v = 1$			$d = 3, \lambda_v = 0,5$			$d = 4, \lambda_v = 1$		
	θ	p	z	θ	p	z	θ	p	z	θ	p	z
20	0.224	0.361	1.918	0.091	0.203	3.982	0.182	0.367	3.172	0.05	0.139	6.025
50	0.141	0.246	2.261	0.044	0.103	4.482	0.087	0.197	4.018	0.02	0.058	6.592
100	0.1	0.181	2.456	0.025	0.061	4.694	0.05	0.118	4.402	0.01	0.029	6.793
10^3	0.032	0.061	2.816	0.004	0.010	4.951	0.008	0.020	4.901	0.001	0.003	6.979
10^4	0.010	0.020	2.941	0.001	0.003	4.992	0.001	0.003	4.984	-	-	6.988

As a short comment to the Table 1, let's note, that at the chain length propagation an average variance of a step of the *SARW* trajectories is increased, in the limit $N \to \infty$ tending to the value 2−1; deformation of the *Flory* ball for example, under converting of the polymeric chain of the ideal solution into the real one or under the action of the external forces, in particular of the shear ones under the gradient rate of the hydrodynamic flow, increases of θ and p and decreases z, sharply decreasing the number of the configurations realizing the presented conformational state.

As it was note earlier, an average probability p to discover of cell occupied due to local character of the *SARW* trajectories is more than the average occupation θ of the cell into the conformational space of a polymeric chain. This permits us like to the determination of θ accordingly to Eq. (22), to express of p via the average configurational volume V_c which consists of a part of the conformational volume $V = R_f^d \lambda_v$:

$$p = a^d N / V_c \qquad (29)$$

Comparing the Eqs. (22) and (29), we will obtain

$$V_c / V = \theta / p \qquad (30)$$

In Table 2 there are calculated values of an average part of the configurational volume on conformational one under different variants.

TABLE 2 Calculated values of an average part of the configurational volume on conformational one under different variants.

N	$V_c/V = \theta/p$			
	$d = 2,\ \lambda_v = 1$	$d = 3,\ \lambda_v = 1$	$d = 3,\ \lambda_v = 0{,}5$	$d = 4,\ \lambda_v = 1$
20	0.620	0.448	0.496	0.359
50	0.574	0.424	0.444	0.344
100	0.551	0.413	0.425	0.339
1000	0.516	0.402	0.400	0.333
10,000	0.500	0.400	0.400	0.333

As we can see, the configurational volume occupies a great part of the conformational volume, testifying to "smeared" *SARW* trajectory into the space of a walk. At the length of a chain propagation the part V_c/V is decreased and in a range $N \rightarrow \infty$ is stabilized by the ratio:

$$V_c/V = 2(\,d+2)$$
(31)

3.3.1.2 CONCENTRATED SOLUTIONS AND MELTS

In accordance with the conclusion done from the Eq. (9), which determines an average variance of a step of the *SARW* trajectories for the linear polymeric chain via average probability to discover the occupied cell, it kept true for any polymeric chain into the concentrated solutions and melts, but at this the p value should be additionally depended on the concentration of polymer. Let us show also, that the main Eqs. (25) and (28) for the concentrated solutions are kept in the previous form (for short the term "melt" will be used as the need arises).

The *SARW* statistics [13] of the linear polymeric chains into the concentrated solutions is based on the notion of m-ball of the intertwined between themselves linear polymeric chains by the same length N with the conformational radius R_m:

$$R_m = R_f m^{1(d+2)} \tag{32}$$

Number m of the chains into the m-ball depends on the concentration of polymer in the solution:

$$m^{2(d+2)} = \rho / \rho^* \tag{33}$$

where: ρ is the density, and ρ^* is the critical density of the solution upon polymer, to which corresponds the start of the *Flory* balls conformational volumes overlapping. It is determined by the expression:

$$\rho^* = M_0 N / N_A R_f^d, \tag{34}$$

in which M_0 is the molar mass of the link of a chain; N_A is the *Avogadro* number.

By introducing the density ρ_0 into a volume of the monomeric link a^d via the ratio

$$\rho_0 = M_0 / N_A a^d, \tag{35}$$

the Eq. (34) can be rewritten in a form:

$$\rho^* = \rho_0 N^{2(1-d)(d+2)} \tag{36}$$

An average occupancy of the cell θ into the conformational volume $R_m^d \lambda_v$ of m-ball can be determined standardly

$$\theta = a^d m N / R_m^d \lambda_v \tag{37}$$

Here, as before, the λ_v parameter is the multiplicity of the volumetric deformation of the m-ball; into the ideal solution and melt $\lambda_v = 1$, into the real concentrated solutions $\lambda_v < 1$.

With taking into account of the previous Eqs. (32)-(36) It's follows from the Eq. (37)

$$\theta = \frac{\rho}{\rho_0} / \lambda_v \qquad (38)$$

So, an average occupancy of the cell θ into the concentrated solutions is the linear function of the concentration of polymer and should be weakly depend on the length of a chain only via parameter λ_v, which can slightly decreased in the real solutions at the N propagation [14].

The density of probability $\omega(\lambda)$ for any linear polymeric chain into m-ball is described by the expression like to Eq. (16), but via the conformational radius of the m-ball:

$$\omega(\lambda) = \exp\left\{-\frac{d+2}{2}\left(\frac{R_m}{\sigma_0}\right)^2 / \lambda_v\right\} \qquad (39)$$

Here as same as earlier, $\sigma_0^2 = aN$; that is why from the determination of R_m accordingly to Eq. (32) and following Eqs. (33)-(36) follows:

$$(R_m / \sigma_0)^2 = \frac{\rho}{\rho_0} N, \qquad (40)$$

This permits to rewrite the Eq. (39) in the next form:

$$\omega(\lambda) = \exp\left\{-\frac{d+2}{2}\frac{\rho}{\rho_0} N / \lambda_v\right\}. \qquad (41)$$

By substituting of the Eq. (41) into determination Eq. (18) the numbers L of the *SARW* trajectories for any linear polymeric chain into m-ball, we will obtain:

$$L = \frac{2d}{2d-1}\left[(2d-1)\exp\left\{-\frac{d+2}{2}\frac{\rho}{\rho_0} / \lambda_v\right\}\right]^N. \qquad (42)$$

This implies the expression for average variance of the step of *SARW* trajectories of the linear polymeric chain into concentrated solutions and melts:

$$z = (2d - 1)\exp\left\{-\frac{d+2}{2}\frac{p}{p_0} / \lambda_v\right\},\tag{43}$$

which in turn with taken into account of the Eq. (38) takes a form

$$z = (2d - 1)\exp\left\{-\frac{d+2}{2}\theta\right\}\tag{44}$$

So, the difference between z for diluted and concentrated solutions is determined by the expressions of numerical estimation of θ. Therefore, the relationship between p and θ for concentrated solutions is kept in the previous form (Eq. 25), which for more convenience can be rewritten as follows

$$p = 1 - \exp\left\{-\frac{d+2}{2}\theta\right\}\tag{45}$$

Let us demonstrate as the illustration in Table 3 the numerical estimations of θ, p and z, and also the ratios $\theta / p = V_c / V_m$ in which V_m is the conformational volume, and V_c is the configurational volume of polymeric chain into m-ball, at different N and p. Calculated done for variant $d = 3$, $\lambda_v = 1$ on example of polystyrene for which $M_0 = 104.15$ g/mole, $a = 1.86 \times 10^{-10}$ m; that is why in accordance with Eqs. (35) and (36) we have $p_0 = 26.9 \times 10^6$ g/mole and $p^* = 0.6757$ and 0.1071 g/mole at $N = 10^2$ and $N = 10^3$, respectively.

TABLE 3. Numerical estimations.

	$N = 100$					$N = 1000$				
$\dfrac{p}{p^*}$	1	1.1	1.2	1.3	1.554	1	2	3	4	9.804
$\theta = p / p_0$	0.025	0.028	0.030	0.033	0.039	0.004	0.008	0.012	0.016	0.039
p	0.061	0.067	0.073	0.078	0.093	0.001	0.020	0.029	0.039	0.093
z	4.696	4.666	4.638	4.608	4.535	4.951	4.901	4.853	4.805	4.535
θ / p	0.413	0.414	0.415	0.416	0.420	0.402	0.404	0.406	0.408	0.420

The values $p / p^* = 1.554$ and $= 9.804$ correspond to the polystyrene melts.

As we can see from the Table 3, at the chosen values N the ratio θ/p is near to the limited one $2/(d+2) = 0.4$. Thus, even into the concentrated solutions the configurational volume, which is an average volume of the SARW trajectories, consists of the great part of the conformational volume that assumes a strong interweaving of the polymeric chains into m-ball. At the polymer concentration increasing at $\rho/\rho^* > 1$ an average variance of the SARW trajectory is visibly decreased that corresponds to the sharp decreasing of the number of configurations L, realizing the conformational state of the polymeric chain into m-ball. An Independence of the presented calculated parameters on the length of a chain is good shown upon their similar values for the melts.

3.3.2 POLYMERIC STARS

3.3.2.1 DILUTED SOLUTIONS

Let the polymeric star consists from the s rays equal to N length. For any pair of rays forming the linear chain by $2N$ length, the SARW statistics [15] determines the density of distribution by the expression:

$$\omega(\lambda) = \exp\left\{-\frac{d+2}{2}\left(\frac{R_s}{\sigma_0}\right)^2 / \lambda_v\right\}, \qquad (46)$$

in which $\sigma_0^2 = a^2 2N$, and R_s is the conformational radius of any un-defined chosen pair of rays, determining also the general conformational volume $R_s^d \lambda_v$ of the polymeric star:

$$R_s = a(2N)^{3(d+2)}(s/2)^{1(d+2)}. \qquad (47)$$

It follows from this

$$\omega(\lambda) = \exp\left\{-\frac{d+2}{2}(2N)^{(4-d)(d+2)}(s/2)^{2(d+2)} / \lambda_v\right\} \qquad (48)$$

By substituting of this expression into determination of L accordingly to Eq. (18), we will obtain

$$L = \frac{2d}{2d-1}\left[(2d-1)\exp\left\{-\frac{d+2}{2}(2N)^{2(1-d)(d+2)}(s/2)^{2(d+2)}/\lambda_v\right\}\right]^N \qquad (49)$$

that gives the possibility to express an average variance of a step of the *SARW* trajectories for any pair of rays of the polymeric star:

$$z = (2d-1)\exp\left\{-\frac{d+2}{2}(2N)^{2(1-d)(d+2)}(s/2)^{2(d+2)}/\lambda_v\right\} \qquad (50)$$

An average occupancy of a cell into conformational volume of the polymeric star we find from the expression:

$$\theta = \frac{a^d sN}{R_s^d \lambda_v}, \qquad (51)$$

which can be rewritten in a form

$$\theta = (2N)^{2(1-d)(d+2)}(s/2)^{2(d+2)}/\lambda_v \qquad (52)$$

Comparing Eqs. (50) and (52), we have again

$$z = (2d-1)\exp\left\{-\frac{d+2}{2}\theta\right\}. \qquad (53)$$

Therefore, the Eq. (24) is kept also for the polymeric star.

The numbers of configurations for pair of rays forming the linear chain by $2N$ length, and for the whole polymeric star taking into account that the number of the independent pairs consisting of s rays equal to $s(s-1)/2$, will be equal correspondingly:

$$L_{2N} = z^{2N}, \; L_{sN} = z^{s(s-1)N}, \qquad (54)$$

For demonstration of the dependence of θ, p and z parameters on the number of rays s in polymeric star in Table 4 presented their values at $2N = 100$, $d = 3$, $\lambda_v = 1$.

TABLE 4   Selected Parameters Dependencies.

s	2	3	6	9	12	15	18	21
θ	0.0251	0.0295	0.0390	0.0458	0.0514	0.0562	0.0605	0.0643
p	0.0608	0.0712	0.0929	0.1083	0.1207	0.1312	0.1403	0.1486
z	4.6956	4.6441	4.5357	4.4585	4.3966	4.3442	4.2982	4.2571
θ/p	0.413	0.415	0.420	0.423	0.426	0.429	0.431	0.433

3.3.2.2   CONCENTRATED SOLUTIONS AND MELTS

SARW statistics of the polymeric stars into the concentrated solutions, as same as the linear chains, is based on the conception of the m-ball of intertwining between themselves polymeric stars. For any pair of rays into undefined star of the m-ball the density of distribution is as follow:

$$\omega(\lambda) = \exp\left\{-\frac{d+2}{2}\left(\frac{R_{ms}}{\sigma_0}\right)^2 / \lambda_v\right\} \tag{55}$$

Here $\sigma_0^2 = a^2 2N$, and R_{ms} is the conformational radius of the m-ball of polymeric stars:

$$R_{ms} = a(2N)^{3/(d+2)}(ms/2)^{1/(d+2)} \tag{56}$$

From the determination

$$\theta = a^d msN / R_{ms}^d \lambda_v \tag{57}$$

with taking into account of the ratios $m^{2(d+2)} = \rho/\rho^*$, $\rho^* = \rho_0(2N)^{2(1-d)/(d+2)}(s/2)^{2/(d+2)}$ it can be written

$$\theta = \frac{\rho}{\rho_0} / \lambda_v \tag{58}$$

So, into the concentrated solutions of polymeric starts the value θ does not depend on the length and the number of rays, but only on the concentration of polymer into solution.

Next, using the developed algorithm and the Eqs. (55)-(58), the standard expression for z type Eqs. (24) (44) and (53) can be again obtained.

3.4 CONCLUSION

The number of configurations L for linear polymeric chain in $d \geq 2$ measured lattice space accurate within multipliers $2d/z$ or $2d/(2d-1)$, neared to unit, is unambiguously determined via the average variance of the step z of the *SARW* trajectories:

$$L \cong z^N \tag{59}$$

The probabilistic analysis of the *SARW* trajectories determines z as the mathematical expectation of the number of free among $2d-1$ neighboring cells via average upon the all *SARW* trajectories probability p to discover the occupied cell. It leads to the expression:

$$z = (2d - 1)(1 - p), \tag{60}$$

in which, however, the value p is kept indeterminate.

SARW statistics, which considers the conformation of polymeric chain as the result of the statistical average upon the all its possible configurations with taking into account of the probability of their realization, leads to the ratio:

$$z = (2d - 1) \exp\left\{-\frac{d+2}{2}\theta\right\}, \tag{61}$$

in which θ is an average upon the conformational volume occupancy of cell or probability to discover the cell occupied into the conformational volume.

From the comparison of Eqs. (60) and (61) the next relationship follows:

$$p = 1 - \exp\left\{-\frac{d+2}{2}\theta\right\} \tag{62}$$

On the basis of values p and θ it can be determined the ratio of the average local volume of configuration V_c to the conformational volume V of polymeric chain:

$$V_c/V = \theta/p \tag{63}$$

At the N increasing this ratio is tended to its limit

$$\theta/p \rightarrow 2/(d+2) \ \mathrm{N} \rightarrow \infty \tag{64}$$

pointing on the great smeared upon the average of the *SARW* trajectory into conformational volume of the polymeric chain.

The Eqs. (59)-(63) are universal in sense that they are true for any linear polymeric chain, including the superposed from undefined pair of rays of polymeric star, in diluted and concentrated, ideal and real solutions. Into diluted solutions θ, p and z depend on the length of a chain, and correspondingly on length and number of rays in polymeric star; in concentrated solutions these parameters are function only on the concentration of polymer.

KEYWORDS

- **configuration**
- **conformation**
- **linear polymeric chains**
- **polymer stars**
- **SARW statistics**

REFERENCES

1. *de* Gennes P. G., Skaling Concepts in Polymer Physics, *"Cornell University Press"* (1979).
2. *des* Cloizeaux J., Jannink G. Polymers in Solution, *"Clarendon Press"*, Oxford (1990).
3. Watts, M. G. *"J. Phys. A: Math. Gen."*, **8**, 61 (1975).
4. Guida, R., Zinn-Justin, J. *"J. Phys. A."*, **31**, 8104 (1998).
5. Duplantier, B. *"Phys. Rev. Lett."*, **57**, 941 (1986).
6. Duplantier, B. *"J. Stat. Phys."*, **54**, 581 (1989).
7. Ferber, C., Holovatch, Yu. *"Cond. Matt. Phys."*, **5** (1), 117 (2002).
8. Ohno, K. *"Cond. Matt. Phys."*, **5** (1), 29, 15 (2002).
9. Batoulis, J., Kremer, K. *"Macromolecules"*, **22** (11), 4277 (1989).
10. Barret, A. J., Tremain, D. L. *"Macromolecules"*, **20**, 1687 (1987).
11. Medvedevskikh, Yu. G., Kytsya, A. R., Bazylyak, L. I., Zaikov, G. E. In: *"Conformation of Macromolecules"* / Editors: Yu. G. Medvedevskikh, et al. Nova Science Publishers, Inc. 173–192 (2007).
12. Medvedevskikh, Yu. G. *"Cond. Matt. Phys."*, **4** (2) (26), 209, 219 (2001).
13. Medvedevskikh, Yu. G. *"J. Appl. Pol. Sci."*, **109** (4), 2472 (2008).
14. Medvedevskikh, Yu. G., Kytsya, A. R., Bazylyak, L. I. In: *"Conformation of Macromolecules"* Editors: Yu. G. Medvedevskikh et al. Nova Science Publishers, Inc., 35-53 (2007).
15. Medvedevskikh, Yu. G., Bazylyak, L. I., Zaikov G. E. Polymeric stars in self-avoiding random walks statistics (*see* Chapter 2 in this book).

CHAPTER 4

CONFORMATION OF LINEAR POLYMERIC CHAINS AT THE INTERFACE LAYER LIQUID / SOLID AND ADSORPTION ISOTHERM

YU. G. MEDVEDEVSKIKH and G. I. KHOVANETS'

CONTENTS

SUMMARY

A conception about volumetric form of the adsorption of polymeric chains has been used accordingly to which their "anchor" fit on the surface of the adsorbent is realized by a little number of the end links forming the Langmuir connection with the active centers of the adsorbent. On a basis of the self-avoiding random walks statistics developed earlier for the solutions, the expressions for the conformational radiuses and for the free energies of polymeric chains in adsorptive diluted and concentrated layers were obtained. Free energy of the conformation included in the determination of the chemical potential of the polymeric chain in solution and adsorption layer. On its basis it was obtained a general expression for the adsorption isotherm, which contains a change of the free conformation energy under transition of a chain from the solution into the adsorptive layer. The partial variants of the general adsorption isotherm were analyzed and it was shown that they give a true description of typical experimentally observed adsorption isotherms.

4.1 INTRODUCTION

An adsorption of the polymeric molecules at an interface layer of liquid and solid phases essentially changes energy and the entropy of the interfacial interaction. That is why it widely used in many practical applications, in particular, for the improvement of the properties of composite materials by functionalization of the filler's surface, for the stabilization of nanoparticles in solutions [1–3] and for the giving of property of the biological compatibility to the (bio)artificial organs in medicine [4]. The scientific interest to the problem of the polymers adsorption is determined also by the variety of factors having an influence on its equilibrium value and dynamics, namely: an availability, number and nature of the functional groups in molecule of polymer, its length N and conformation, thermodynamical quality of the solvent and nature of the adsorptive (active) centers of the surface of solid.

   In accordance with the earlier model [5–9] of the separate molecule of polymer ad-sorption, on the surface of the adsorbent the polymeric chains having the conformation of balls in solution are straighten up and form the plate although "diffusive", that is with loops and tails layer by h thickness, which is considerably less than the conformational radius R_f of the *Flory* ball in solution: $h < R_f$. An essential argument for benefit of this model is fact, that it supposes a considerable energy of the *Langmuir* interaction of active centers of the adsorbent with more number of polymer's links. It seems, that the investigations [10–11] with the use of the *IR*-spectroscopy confirm of this fact indicating on the great part (from 0.2 till ~ 1) connected with the active centers of monomeric links of a chain.

However, the modern experimental methods, in particular, an ellipsometry and the neutron reflection [12], give the values of the adsorption layer thickness h neared to the sizes of the *Flory* ball in solution, $h \approx R_f$, that points on the volumetric character of the polymer adsorption but not on the plate one.

These and others experimental facts have been theoretically analyzed with the use of the methods of the self-consistent field and scaling [13–17]. The results of an analysis can be lined in the following simplified model: in the flocculent adsorption layer (a distance between the centers of the adsorptive molecules $l \geq 2R_f$, the polymeric chain is in practically the same conformational state as in the solution; in dripless adsorptive layer ($l < 2R_f$) an interaction between the adsorbed chains compresses the polymeric balls in the adsorption plate and stretches them in a form of the chain by blobs [16], cylinders [13] or rotation ellipsoids [18] along the normal to the surface.

Presented model doesn't not take into account that the interaction of polymeric chains in the dripless adsorptive layer can leads to the overlapping of their conformational volumes with the formation of physical network of the intertwined between themselves polymeric chains.

A model of flocculent and dripless adsorptive layers also is good agreed with the experimental ellipsometric data on kinetics of adsorption of different upon nature and length polymeric chains and also on number and distribution in them of the functional groups [19–23]. It is follows from these references that the kinetic curves of adsorption of the polymeric chains from the diluted solutions consist of two sections: the first is starting – quick and short with the characteristic time by 10^2 s order and the second one (final) – slow and long with the characteristic time

~ 10^3 s. It was notified in Ref. [*19–23*], that the kinetic constant of quick and slow process, the capacity of flocculent and dripless adsorptive layers, practically (that is via the measurements error limits), do not depend on the nature, number and type of the functional groups location in polymeric chain, but depend considerably on its length and consequently, on its conformational state.

Thereby, a question about the conformational state of the polymeric chain in the adsorptive layer consists in the following: what is the more advantageous from the point of view of the free energy *(Helmholtz)*: **the plate** ($h < R_f$) **form of an adsorption** (to which a high energy of the *Langmuir* interaction "adsorbate–adsorbent" and low entropy of the conformation of the adsorbed polymeric chain correspond), *or the volumetric* ($h \approx R_f$) *form of an adsorption* (at which a loss of the part of energy of *Langmuir* interaction can be compensated by a high entropy of the conformational state of the polymeric chain in the adsorption layer).

In the presented paper we will be started from the imagination about the volumetric form of the polymeric chains adsorption assuming that the "anchor" fit of the polymeric chain on the surface of adsorbent is realized via little number of z ($z < N$, where N is a general number of links of the polymeric chain) of the end links forming the *Langmuir* connection with the active centers of adsorbent. In spite of fact, that the presented model, as it was mentioned earlier, in detail was analyzed with the use of the methods of self-consistent field and scaling, but it was not obtained its thermodynamical evolution.

Here the thermodynamical analysis of the polymeric chains adsorption will be based on the full taking into account of free energy of conformation of polymeric chains in solution and adsorptive layer. The notions "flocculent and dripless adsorptive layers" will be substituted on thermodynamically more determined notions of diluted and concentrated adsorption layers.

4.2 CONFORMATION AND FREE ENERGY OF CONFORMATION OF THE LINEAR POLYMERIC CHAINS IN ADSORPTIVE LAYER

4.2.1 DILUTED ADSORPTIVE LAYER

Equilibrium conformational state and its free energy for linear polymeric chains in adsorptive layer let's determine starting from the following characteristics of the diluted solution in which $r \leq r^*$, where r and r^* are density and critic density of the solution upon polymer, corresponding to the beginning of the polymeric chins conformational volumes overlapping.

Self-avoiding random walks statistics *(SARWS)* [24] for diluted solution determines the distribution density $w\,(x, N)$ of the end of a chain from its beginning in d-measured space via expression:

$$\omega(x,N) = \exp\left\{-\frac{a^d N^2}{\prod_i x_i} - \frac{1}{2}\sum_i \frac{x_i^2}{\sigma_i^2}\right\}, \; i = 1, d \tag{1}$$

where: a is a linear size of the link of a chain, x_i is the displacement of the end of a chain from its beginning along the i direction, s_i is the mean square deviations of the *Gaussian* part of the distribution (Eq. 1), connected via the ratio:

$$\sum_i \sigma_i^2 = a^2 N d, \; i = 1, d \tag{2}$$

Among the all-possible states of the polymeric chain let's choice the most probable or thermodynamically equilibrium in accordance with the condition:

$$\partial\omega(\delta,N)/\partial x_i = 0 \; \text{ at } \; x_i = X_i \tag{3}$$

This leads to the determination of the equilibrium semi-axes of the conformational ellipsoid of the polymeric chain in the real solution:

$$X_i = \sigma_i\left(a^d N^2 \Big/ \prod_i \sigma_i\right)^{1/(d+2)} \tag{4}$$

In the ideal solution the all directions of the chain's end walk are equiprobable, that leads to the condition resulting from the Eq. (2):

$$\sigma_i^2 = \sigma_0^2 = a^2 N \qquad (5)$$

By combining the Eqs. (4) and (5) we find the conformational radius of the polymeric chain in the ideal solution or the radius of the un-deformed *Flory* ball:

$$R_f = a N^{3/(d+2)} \qquad (6)$$

In the real solution the Flory ball is deformed into the rotation ellipsoid, compressed or elongated along the axe connecting the beginning and the end of a chain. The semi-axes X_i of the conformational ellipsoid can be expressed via the R_f and vie the multiplicities l_i of the linear deformation:

$$X_i = R_f \lambda_i \qquad (7)$$

The volumetric deformation we determine via the ratio:

$$\prod_i X_i = R_f^d \prod_i \lambda_i \qquad (8)$$

in which

$$\prod_i \lambda_i = \lambda_v$$

is the multiplicity of the volumetric deformation.

The multiplicities of the linear and volumetric deformation are not un-defined and are connected via the ratio [24]:

$$\sum_i \lambda_i^2 = d/\lambda_v \qquad (9)$$

As the analysis in Ref. [24] shows, at any deformations of the *Flory* ball the conformational volume is decreased $\prod_i X_i < R_f^d$, that is why $\lambda_v < 1$. Thereby, in general case $\lambda_v \leq 1$, and the sign "=" corresponds to the ideal solution, and the sign "<" corresponds to the real solution.

An entropy S and free energy F of the equilibrium conformational state of polymeric chain let's determine via the expressions:

$$S = k \ln \omega(x, N), \; F = -kT \ln \omega(x, N) \text{ at } x_i = X_i \qquad (10)$$

It is follows from the determination of Eq. (10) and from the Eqs. (1) and (7) for the real solution:

$$F = \frac{d+2}{2} kT \left(\frac{R_f}{\sigma_0} \right)^2 \Big/ \lambda_v , \qquad (11)$$

for the ideal one

$$F^0 = \frac{d+2}{2} kT \left(\frac{R_f}{\sigma_0} \right)^2 \qquad (12)$$

Here

$$\left(R_f / \sigma_0 \right)^2 = N^{(4-d)/(d+2)} \qquad (13)$$

Accordingly to the formulated above starting positions the distribution density ω_s (x, N) of the end of a linear polymeric chain from its beginning in the adsorbed layer can be written in the same form (Eq. 1), but with the correction that due to the presence of reflecting surface a half of the d-measured volume becomes inaccessible for the displacements x_i of the end of a chain. That is why an expression for ω_s (x, N) we will obtained from the Eq. (1) by the substitution $\prod\limits_i x_i / 2$ instead of $\prod\limits_i x_i$. So, we will obtain:

$$\omega_s(x, N) = \exp\left\{ -\frac{2a^d N^2}{\prod\limits_i x_i} - \frac{1}{2} \sum\limits_i \frac{x_i^2}{\sigma_i^2} \right\} \qquad (14)$$

Using the conditions of Eq. (5) we find the equilibrium values X_{si} of the semi-axes of conformational ellipsoid of polymeric chain in the real adsorption layer:

$$X_{si} = \sigma_i \left(2a^d N \Big/ \prod\limits_i \sigma_i \right)^{1/(d+2)} \qquad (15)$$

For the ideal adsorptive layer accordingly to the Eq. (5) we have:

$$R_s = aN^{3/(d+2)}2^{1/(d+2)} \tag{16}$$

So, in adsorption layer the conformational radius of the polymeric chain is even more, than in the solution: $R_s = R_f 2^{1/(d+2)}$. This is connected with fact that the demand of the absence of the self-avoiding walks trajectories at the inaccessibility of the volume under the reflecting surface leads to the capture of more volume above the reflecting surface.

We will obtain the expressions for the free energy F_s equilibrium to the conformation of polymeric chain in the real adsorption layer from the determination of Eq. (10) with taking into account of Eqs. (7) (8) (15) and (16):

$$F_s = \frac{d+2}{2}kT\left(\frac{R_s}{\sigma_0}\right)^2 \bigg/ \lambda_{sv} \tag{17}$$

and in the ideal one

$$F_s^0 = \frac{d+2}{2}kT\left(\frac{R_s}{\sigma_0}\right)^2 \tag{18}$$

Here $\lambda_{sv} = \prod_i \lambda_{si}$ is the multiplicity of the volumetric deformation of conformational volume in real adsorbed layer connected with the multiplicities of the linear deformations λ_{si} via the ratio similar to Eq. (9).

It is follows from the Eqs. (6) and (9):

$$\left(R_s/\sigma_0\right)^2 = \left(R_f/\sigma_0\right)^2 2^{2/(d+2)} \tag{19}$$

That is why it can be written instead of the Eq. (18) with taking into account of Eq. (12):

$$F_s^0 = F^0 2^{2/(d+2)} \tag{20}$$

So, at the transition of a polymeric chain from the ideal diluted solution into the ideal one-adsorption layer the change of the free energy of conformation $\Delta F = F_s^0 - F^0$ will be positive and will be as follow:

$$\Delta F = \frac{d+2}{2} kT \left(R_f / \sigma_0 \right)^2 \left(2^{2/(d+2)} - 1 \right) \tag{21}$$

4.2.2 CONCENTRATED ADSORPTION LAYER

As same as earlier, the discussion let's start from the concentrated solution, in which accordingly to condition $\rho \geq \rho^*$ the conformational volumes of the polymeric chains are overlapped. That is why, an object of the *SARW* statistics in this case is some volume of the d-measured space, in which there are m intertwined between themselves polymeric chains. It is follows from the analysis [25] that the distribution density of the end relatively the beginning for any undefined chosen from the m-ball of polymeric chain will be determined by the expression:

$$\omega(x,m,N) = \exp \left\{ -\frac{a^d m N^2}{\prod_i x_i} - \frac{1}{2} \sum_i \frac{x_i^2}{\sigma_i^2} \right\} \tag{22}$$

We find from this accordingly to the same condition (Eq. 3) the equilibrium semi-axes of the conformational ellipsoid in the real concentrated solution:

$$X_{mi} = \sigma_i \left(a^d m N^2 / \prod_i \sigma_i \right)^{1/(d+2)} \tag{23}$$

and in accordance with the condition Eq. (5) the conformational radius of any undefined chain and m-ball in wholly in the ideal concentrated solution will be equal:

$$R_m = a N^{3/(d+2)} m^{1/(d+2)} \tag{24}$$

At this, it can be again written

$$X_{mi} = R_m \lambda_i \tag{25}$$

$$\prod_i X_{mi} = R_m^d \lambda_v \tag{26}$$

where $\lambda_v = \prod_i \lambda_i$ is the multiplicity of the volumetric deformation of the m-ball. The values λ_i and λ_v are connected as before by the Eq. (9).

From the determination of Eq. (10) after the substitution in Eq. (22) of the equilibrium values $x_i = X_{mi}$ accordingly to Eq. (23) with taking into account of Eqs. (25) and (26) we will obtain the expression for free energy of conformation F_m for any undefined chain in m-ball in concentrated solution: real

$$F_m = \frac{d+2}{2} kT \left(\frac{R_m}{\sigma_0} \right)^2 \Big/ \lambda_v \tag{27}$$

and ideal

$$F_m^0 = \frac{d+2}{2} kT \left(\frac{R_m}{\sigma_0} \right)^2 \tag{28}$$

Here accordingly to Eq. (24)

$$(R_m/\sigma_0)^2 = (R_f/\sigma_0)^2 m^{2/(d+2)} \tag{29}$$

Let's transfer the obtained ratios into concentrated adsorptive layer. From the considerations above presented a distribution density in concentrated adsorption layer takes a form:

$$\omega_s(x, m_s, N) = \exp\left\{ -\frac{a^d 2 m_s N^2}{\prod_i x_i} - \frac{1}{2} \sum_i \frac{x_i^2}{\sigma_i^2} \right\} \tag{30}$$

Therefore, we have for the real concentrated adsorptive layer:

$$X_{smi} = \sigma_i \left(a^d 2 m_s N^2 \Big/ \prod_i \sigma_i \right)^{1/(d+2)}, \tag{31}$$

for ideal

$$R_{sm} = aN^{3/(d+2)} (2m_s)^{1/(d+2)} \tag{32}$$

Here m_s is a number of the chains in m_s-ball of the adsorption layer.

Free energy of conformation for any undefined chain in the m_s-ball in concentrated adsorbed layer is determined by the expressions: for real

$$F_{sm} = \frac{d+2}{2} kT \left(R_{sm}/\sigma_0 \right)^2 \Big/ \lambda_{sv} \tag{33}$$

for ideal

$$F_{sm}^0 = \frac{d+2}{2} kT \left(R_{sm}/\sigma_0 \right)^2 \tag{34}$$

where again $\lambda_{sv} = \prod_i \lambda_{si}$ is the multiplicity of the volumetric deformation of m_s-ball in real concentrated adsorbed layer.

From the comparison of Eqs. (32) and (6) follows

$$\left(R_{sm}/\sigma_0 \right)^2 = \left(R_f/\sigma_0 \right)^2 \left(2m_s \right)^{2/(d+2)} \tag{35}$$

That is why it can be also written:

$$F_{sm}^0 = F^0 \left(2m_s \right)^{2/(d+2)} \tag{36}$$

So, at the transfer of polymeric chain from the ideal concentrated solution into the ideal concentrated adsorption layer a free energy of conformation increases on the value:

$$\Delta F = \frac{d+2}{2} kT \left(R_f/\sigma_0 \right)^2 \left[\left(2m_s \right)^{2/(d+2)} - m^{2/(d+2)} \right] \tag{37}$$

The values m and m_s depend on the concentration of polymer in solution and in the adsorption layer correspondingly. Let's determine of their dependence on a density upon polymer of the solution r and adsorption layer ρ_s. In concentrated solution and in adsorption layer the density upon polymer in the conformation volumes m- and m_s-ball is the same as in the whole volume of the solution and in the whole volume of the adsorption layer. That is why:

$$\rho = M_0 mN / N_A R_m^d \tag{38}$$

$$\rho_s = M_0 2 m_s N / N_A R_{sm}^d \qquad (39)$$

Here M_0 is a molar mass of the link of a chain. In Eq. (39) the number 2 was appeared in order to take into account that at presence of the reflecting surface, the conformational volume of the polymeric chain in the adsorption layer is equal to $R_{sm}^d / 2$.

At the value $m = 1$ we have $R_m = R_f$, and $\rho = \rho^*$ is the critic concentration of the polymeric chains in the solution corresponding to the beginning of their conformational volumes overlapping:

$$\rho^* = M_0 N / N_A R_f^d \qquad (40)$$

Similarly, at $m_s = 1$ we have $R_{sm} = R_s$, and $\rho_s = \rho_s^*$, where

$$\rho_s^* = M_0 2 N / N_A R_s^d \qquad (41)$$

From the comparison of Eqs. (38) and (40) (39) and (41) with taking into account of the expressions for R_f, R_m, R_s and R_{sm} accordingly to Eqs. (6) (24) (16) and (32) we find

$$m^{2/(d+2)} = \rho / \rho^* \qquad (42)$$

$$m_s^{2/(d+2)} = \rho_s / \rho_s^*. \qquad (43)$$

Critical concentrations of the beginning of the polymeric chains conformational volumes overlapping in solution r^* and in adsorption layer ρ_s^* can be expressed in the form of clear dependence on the length of a chain:

$$\rho^* = \rho_0 N^{2(1-d)/(d+2)} \qquad (44)$$

$$\rho_s^* = \rho_0 N^{2(1-d)/(d+2)} 2^{2/(d+2)} \qquad (45)$$

where

$$\rho_0 = M_0/N_A a^d \tag{46}$$

is per sense the density in the volume of the monomeric link.

4.3 ADSORPTION ISOTHERM OF THE POLYMERIC CHAINS

4.3.1 OVERALL VIEW OF THE ADSORPTION ISOTHERM

At the adsorption of polymeric chains from the solution the dislodgment of the molecules of a solvent from the active centers of adsorbent takes place; that is why the adsorption process can be considered as quasi-chemical reaction by the following view [3]

$$P + zB_s = P_s + zB \tag{47}$$

Here: P and B, P_s and B_s are a polymer and a molecule of the solvent into solution and into adsorption layer correspondingly.

A condition of the adsorbed equilibrium Eq. (47) has the following standard view

$$\sum_i \upsilon_i \mu_i = 0 \text{,} \tag{48}$$

where μ_i and υ_i are the chemical potential and the stoichiometric coefficient of the i participant in Eq. (47).

Chemical potentials of the polymeric chains in the solution μ_p and in the adsorbed layer μ_{ps} we will determine with taking into account of their free energy of conformation via the following expressions:

$$\mu_p = \mu_p^0 + kT \ln \frac{\rho}{\rho^*} + F \text{,} \tag{49}$$

$$\mu_{ps} = \mu_{ps}^0 + kT \ln \frac{\rho_s}{\rho_s^*} + F_s \text{.} \tag{50}$$

As it can be seen, the values $\mu_p{}^0$ and $\mu_{ps}{}^0$ were determined by the choice of the standard states of the solution at $r = r^*$ and of the adsorption layer at $\rho_s = \rho_s{}^*$.

Next we will be suppose that even in concentrated solution and in adsorption layer accordingly to conditions $r > r^*$ and $\rho_s > \rho_s{}^*$ the molar part of the solvent considerably more than the molar part of the polymeric chains. That is why taking into account also the position $z < N$, it can be assumed, that at the adsorption of polymer the change of the molar part of the molecules of the solvent in solution and in adsorption layer will be little and it can be neglected. That is why, chemical potentials of the solvent in the solution and in the adsorption layer we will determine by the expressions

$$\mu_B \cong \mu_B^0 , \tag{51}$$

$$\mu_B \cong \mu_{Bs}^0 , \tag{52}$$

in which $\mu_B{}^0$ and $\mu_{Bs}{}^0$ are standard chemical potentials of the solvent in solution and in adsorption layer.

Let's rewrite the condition of chemical potential of the equilibrium Eq. (47) as follows:

$$\mu_\beta - \mu_p + z\mu_B^0 - z\mu_{Bs}^0 = 0 . \tag{53}$$

Let's note

$$-\left(\mu_{ps}^0 - \mu_p^0 \right) \big/ z = kT \ln K_p \tag{54}$$

$$-\left(\mu_{Bs}^0 - \mu_B^0 \right) = kT \ln K_B \tag{55}$$

where K_p and K_B are equilibrium constants of the *Langmuir* interaction of polymeric chain and molecule of the solvent with one active center of the adsorbent correspondingly.

It is follows from this that the equilibrium constant K of the displacing adsorption of polymer accordingly to Eq. (47) will be equal to

$$K = \left(K_p / K_B\right)^z \tag{56}$$

Using the determinations Eqs. (54)-(56) and the Eqs. (49) and (52) in (53) we will obtain the overall view of the adsorption isotherm of polymeric chains:

$$\frac{\rho_s}{\rho_s^*} = K\frac{\rho}{\rho^*}\exp\left\{-\Delta F/kT\right\} \tag{57}$$

where $\Delta F = F_s - F$ is a change of a free energy of conformation of polymeric chain at its transfer from the solution into the adsorption layer.

Since the value DF can depends on the ratio of the concentrations of polymer in solution and in adsorption layer at $r^3 \; r^*$ and $\rho_s^{\;3} \; \rho_s^*$, the overall Eq. (Eq. 57) can be filled by the essentially different content.

An analysis of the partial forms of the adsorption isotherms let's carry out under the approximation of the idealness of the solution and of the adsorption layer. Peculiarities of the adsorption polymer from the real solution let's discuss on the qualitative level.

4.3.2 PARTICULAR FORMS OF THE ADSORPTION ISOTHERM

4.3.2.1 HIGH AFFINITY OF POLYMER TO THE ACTIVE CENTERS OF ADSORBENT: K > 1

a) Solution and adsorption layer are diluted accordingly to condition $\rho/\rho^* \leq 1$, $\rho_s/\rho_s^* \leq 1$.

In this variant the change of a free energy of conformation at the transfer of the polymeric chain from the solution in the adsorption layer is determined by the Eq. (21), which let's rewrite taking into account of Eq. (13) for real $d = 3$ space:

$$\Delta F = \frac{5}{2}kTN^{1/5}\left(2^{2/5} - 1\right) \tag{58}$$

Therefore, the adsorption isotherm takes a following view:

$$\frac{\rho_s}{\rho_s^*} = K \frac{\rho}{\rho^*} \exp\left\{-\frac{5}{2} N^{1/5}\left(2^{2/5}-1\right)\right\} \tag{59}$$

This expression describes the initial linear section of the dependence ρ_s / ρ_s^* on r / r^*, as it was shown on the Fig. $1a$ and b at three values of K and two values of N. As we can see, at $K > 1$ the ratio ρ_s / ρ_s^* achieves the value, equal to 1, earlier than the ratio r / r^*. That is why the next section of the adsorption isotherm is characterized by the condition.

b) *Solution is diluted* $\rho/\rho^* \leq 1$, *adsorption layer is concentrated* $\rho_s/\rho_s^* \geq 1$

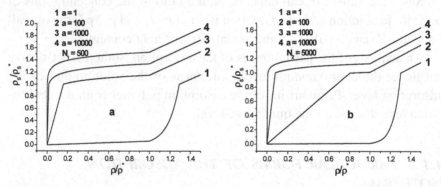

FIGURE 1   Calculated adsorption isotherms of linear polymeric chains at different values of K and N in coordinates $\rho_s / \rho_s^* - r / r^*$.

In the presented variant free energy of conformation of polymeric chain in solution is determined by the Eq. (12), in the adsorption layer by the Eq. (34) or (36) with taking into account of Eq. (42). Combining them we will obtain:

$$\Delta F = \frac{d+2}{2} kT \left(R_f / \sigma_0\right)^2 \left(2^{2/(d+2)} \frac{\rho_s}{\rho_s^*} - 1\right) \tag{60}$$

An adsorption isotherm takes a view:

$$\frac{\rho_s}{\rho_s^*} = K \frac{\rho}{\rho^*} \exp\left\{-\frac{5}{2} N^{1/5}\left(2^{2/5} \frac{\rho_s}{\rho_s^*} - 1\right)\right\} \tag{61}$$

As we can see from the Eqs. (60) and (61), at the transfer of polymeric chain from the diluted solution into the concentrated adsorption layer at ρ_s / ρ_s^* increasing the value ΔF is sharply increased that gives a strong counteracting to the following adsorption effect. That is why the second section of the adsorption isotherm is characterized by a great deceleration of the adsorption value with the growing of polymer concentration in solution and gives the visibility of the going out on plateau (see Fig. 1). Determined on the basis of this quasi-plateau value of adsorption ρ_s is insignificantly more than ρ_s^* and weakly depends on the equilibrium constant K of the displacing adsorption: thus, at the transition from the curve 2 to the curve 4 on Fig. 1 the constant K is increased on two orders, whereas the value of adsorption on $\sim$20%. At the growth of the length of chain N the value of adsorption on the quasi-plateau is decreased.

Presented section of the adsorption isotherm is continued till the value $\rho/\rho^* = 1$. Thus, the next third section of the adsorption isotherm is characterized by the transition of polymeric chains from the concentrated solution in the concentrated adsorption layer.

c) *Concentrated solution and adsorption layer:* $\rho/\rho^* \geq 1, \rho_s/\rho_s^* \geq 1$

In accordance with the previous analysis in the presented variant ΔF is determined by the Eq. (37), which with taking into account of Eqs. (42) and (43) can be written as

$$\Delta F = \frac{d+2}{2} kT \left(R_f/\sigma_0\right)^2 \left(2^{2/(d+2)} \frac{\rho_s}{\rho_s^*} - \frac{\rho}{\rho^*}\right) \tag{62}$$

Using the Eq. (62) and taking into account of Eq. (13) in general Eq. (57), at $d = 3$ we will obtain

$$\frac{\rho_s}{\rho_s^*} = K \frac{\rho}{\rho^*} \exp\left\{-\frac{5}{2}N^{1/5}\left(2^{2/5}\frac{\rho_s}{\rho_s^*} - \frac{\rho}{\rho^*}\right)\right\} \tag{63}$$

Therefore, at transition in a field of the concentrated solution, as it was illustrated by numerous calculations on Fig. 1, the adsorption of polymer is sharply increased forming the third practically linear section of the adsorption isotherm. Comparing of these sections at different equilibrium constants K of the displacement adsorption (curves 2, 3 and 4) it can be note again that the main factor determining the rate of the adsorption change on this section is not the constant K, but the change of a free energy of conformation.

Calculated adsorption isotherms (curves 2 – 4 on Fig. 1) are typical at the considerable affinity of a polymer to the adsorbate. Let's consider, however, the view of the adsorption isotherm with relatively weak affinity of a polymer to the adsorbate.

4.3.2.2 WEAK AFFINITY OF A POLYMER TO ADSORBATE ACCORDINGLY TO CONDITION $K \leq 1$

In this case some previous variants are also possible but in other sequence and in a new quality.

a) Diluted solution and adsorption layer: $\rho/\rho^* \leq 1$, $\rho_s/\rho_s^* \leq 1$.

In presented concentration interval a change of a free energy of conformation ΔF is described by the Eq. (58), and the adsorption isotherm by the Eq. (59). Therefore, presented section of the adsorption isotherm is linear, but with very little angle of inclination (see Fig. 1, curve 1 at $K = 1$), that cannot permit experimentally find it. As we can see, at low values of K the critical concentration of polymer is achieved earlier than in the adsorbed layer. That is why the second section of the adsorption isotherm is characterized by the condition:

b) Concentrated solution and diluted adsorption layer: $\rho/\rho^* \geq 1$, $\rho_s/\rho_s^* \leq 1$.

In this variant an expression for ΔF of the polymer adsorption we will obtain by the combination of the Eqs. (20) (28), and (29) with taking into account of Eq. (42):

$$\Delta F = \frac{d+2}{2} kT \left(R_f/\sigma_0 \right)^2 \left(2^{2/(d+2)} - \frac{\rho}{\rho^*} \right) \tag{64}$$

It's follows from this

$$\frac{\rho_s}{\rho_s^*} = K \frac{\rho}{\rho^*} \exp\left\{ -\frac{5}{2} N^{1/5} \left(2^{2/5} - \frac{\rho}{\rho^*} \right) \right\} \tag{65}$$

Accordingly to Eqs. (64) and (65) in presented concentration interval $\rho/\rho^* \geq 1$ and $\rho_s/\rho_s^* \leq 1$ up to the value $\rho_s/\rho_s^* = 1$ with growth of the concentration of polymer in solution ΔF is decreased, that sharply, practically exponentially increases of the polymer adsorption (see Fig. 1, curve 1). At the achievement of critical concentration of polymer in the adsorption layer the third section of the adsorption isotherm is started corresponding to the condition $\rho/\rho^* \geq 1$, $\rho_s/\rho_s^* \geq 1$. It is wholly identical to the considered above and is described by the same Eq. (63), but at little values of K. At this, comparing the presented on Fig. 1 adsorption isotherms (curves 1–4), it can be again noted that on the third section of the isotherm the rate of the adsorption growth, that is $\partial\rho_s/\partial\rho$, is practically the same in spite of the difference between the constants K on four orders.

4.4   CONCLUSIONS

Usually the experimental values of the adsorption of polymer are described not in the coordinates $\rho_s / \rho_s^* - r / r^*$, but in coordinates $A_s - \rho$, where A_s is a mass of the adsorbed polymer, referred to the unit of the surface of adsorbent. Let's determine the relationship between A_s and ρ_s. Since $A_s =$ mass $/ s$, $\rho_s =$ mass $/ sR_s$, where s is the value of the surface of adsorbent, for diluted adsorption layer we have

$$A_s = \rho_s R_s \text{ at } \rho/\rho_s^* \leq 1 \tag{66}$$

In concentrated adsorption layer

$$A_s = \rho_s R_s \text{ at } \rho/\rho_s^* \geq 1 \tag{67}$$

Critical value of the adsorption A_s^*, corresponding to the beginning of the polymeric chains conformational volumes overlapping in the adsorption layer will be equal:

$$A_s^* = \rho_s^* R_s \tag{68}$$

Using the Eqs. (16) (45), and (46) we will obtain the clear form of the dependence of A_s^* on the properties of polymeric chain:

$$A_s^* = 2^{3/5} \frac{M_0}{N_A a^2} N^{-1/5}$$

(69)

A view of the calculated adsorption isotherms in coordinates $A_s - \rho$ is shown on Fig. 2. They keep the same peculiarities as the presented on Fig. 1.

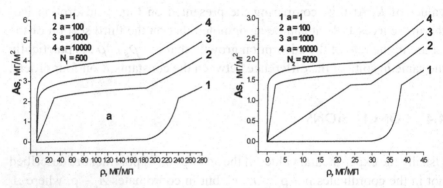

FIGURE 2   Calculated adsorption isotherms for linear polymeric chains at different values of K and N in coordinates $A_s - \rho$.

Comparing of these calculated adsorption isotherms and experimental ones, additionally let's note.

1. At the growth of the length of a chain N, the work, that is the value ΔF of the polymeric chain transfer from the solution into the adsorption layer is increased and the value of adsorption is decreased that is in good agreement with the experimental data, obtained with the use of the ellipsometric methods on the ideally smooth surfaces of the adsorbent. However, the experiments concerning to the adsorption of polymers on the porous adsorbents indicates sometimes on the increasing of the values of adsorption on the quasi-plateau of the isotherm at the growth of the molecular mass of the polymer. From our point of view, this is not the contradiction with the theoretical analysis of the presented work. Evidently that the adsorption of polymeric chains in the pores of adsorbent can characterized by the peculiarities depending on ratio

of the sizes of pores and on the conformational volume of polymeric chains. It is not excluded that in this case the equilibrium constant K of the displacing adsorption of polymer can depend on the molar mass of the polymer at the expense of the additional Langmuir interaction of monomeric links with the active centers of adsorbent on the walls of the pores, that is at the expense of the z increasing. The behavior of polymeric chains in the pores of adsorbent having the size of the pores commensurable with the conformational volume of the chains needs more detailed analysis. However, let us note that the growth of the K on some orders weakly increases the value of the adsorption of polymer on quasi-plateau; that is why the influence of the molar mass of the polymer on values z and correspondingly K, will be also small visualized on the value of the adsorption on quasi-plateau, that is observed in the experiments. An absence of the visible influence of nature of the ω-functional groups of polymer on the value of adsorption on quasi-plateau is also explained by these reasons.

2. The numerous experimental data indicates that at the growth of the polymeric chains hardness and at the impairment of the thermodynamic quality of the solvent the adsorption of polymer is increased. We combine of these variants comparing the adsorption of polymer from the ideal and from the real solutions limiting concretely by the field of the quasi-plateau, in which at $K > 1$ the solution is dilute, and the adsorption layer is concentrated.

In this variant for the estimation of ΔF it is needed to use the Eq. (33) for F_{sm} in a case of the real concentrated adsorption layer and to use the Eq. (11) for F in a case of the real diluted solution. By combining of them we will obtain:

$$\Delta F = \frac{d+2}{2} kT \left(R_f / \sigma_0 \right)^2 \left(2^{2/(d+2)} \frac{\rho_s}{\rho_s^*} \bigg/ \lambda_{sv} - \frac{1}{\lambda_v} \right) \qquad (70)$$

where λ_{sv} and λ_v are multiplicity of the volumetric deformation and conformational volume of polymeric chain in the real adsorption layer and in the solution correspondingly.

In accordance with general Eq. (57), we find

$$\frac{\rho_s}{\rho_s^*} = K \frac{\rho}{\rho^*} \exp\left\{-\frac{5}{2} N^{1/5} \left(2^{2/5} \frac{\rho_s}{\rho_s^*} \Big/ \lambda_{sv} - \frac{1}{\lambda_v}\right)\right\} \qquad (71)$$

At the transfer of the polymeric chain from the ideal into the real solution its conformational volume is deformed with the transformation of the spherical Flory ball into the conformational ellipsoid elongated or flattened along the axis connecting the begin and the end of a chain [26], that leads to decrease of the conformational volume and accordingly to the Eq. (8) to decrease of λ_v: at any deformations of the *Flory* ball λ_v became less than the one. That this why the effects related with the notions "hardness of the polymeric chain" and "the thermodynamic quality of the solvent" can be quantitatively estimated via the multiplicity of the volumetric deformation $\lambda_v \leq 1$. The indicated effects are visualized in the adsorption layer weaker than in the solution: firstly, because the conformational volume in the adsorption layer equal to $R_s^d/2$, is less, than in solution. This increases the elastic properties of the conformational volume of polymeric chain and thereafter increases the deformation work. Moreover, the concentrated adsorption layer corresponding to the quasi-plateau on the adsorption isotherm is more near to the ideal than the diluted real solution. That is why under other equal conditions $\lambda_{sv} > \lambda_v$. This means, that the adsorption of polymer from the real solution is more than from the ideal one.

KEYWORDS

- adsorption isotherm
- conformation
- interface layer
- polymeric chains

REFERENCES

1. Napper D. H. Polymeric stabilization of colloid dispersions, *Academic Press: London* (1975).

2. Adhesion and Adsorption of Polymers / *Ed. by Lee L. H. New York (*1980).
3. Lipatov Yu.. S., Sergeyeva L. M. Adsorption of polymers, *Kiyev: "Naukova dumka"* (1972) *(in Russian)*
4. Ruckenstein E., Chang D.B., *"J. Colloid Interface Sci."*, **123**, p. 170 (1988).
5. Jenkel E., Rumbach B., *"Z. Electrochem."*, **5**, p. 612 (1951).
6. Hoeve C. A. J., *"J. Chem. Phys."*, **44**, p. 1505 (1966).
7. Silberberg A., *"J. Chem. Phys."*, **48**, p. 2835 (1968).
8. Roe R. J., *"J. Chem. Phys."*, **60**, p. 4192 (1974).
9. Schentjens J. M. H. M., Fleer G. J., *"J. Phys. Chem."*, **84**, p. 178 (1980).
10. Kawaguchi M., Yamagiwa S., Takahachi A., Kato T., *"J. Chem. Soc. Faraday Trans."*, **86** (9), p. 1383 (1991).
11. Chornaya V. N., Todosiychuk T. T., Menzheres G. Ya., Konovalyuk V. D., *"Vysoko molekulyarnyye soyedinieniya"*, **51** A (7), p. 1155 (2009) *(in Russian)*.
12. Taunton H. J., Toprakcibylu C., Fetters L. J., Klein J., *"Macromolecules"*, **23**, p. 571 (1990).
13. Alexander S., *"Le J. de Physique"*, **38**, p. 983 (1977).
14. Marques C. M., Joanny J. F., *"Macromolecules"*, **22**, p. 1454 (1989).
15. Ligoure C., Leibler L., *"Le J. de Physique"*, **51**, p. 1313 (1990).
16. De Gennes P. G., *"Macromolecules"*, **13**, p. 1069 (1980).
17. Milner S. T., Witten T. A., Cates M. E., *"Ibid."*, **21**, p. 2610 (1988).
18. Jones R. A. L., Norton L. J., Shull K. L. et al., *"Ibid."*, **25**, p. 2359 (1992).
19. Motschman H., Stamm M., Toprakcioglue G., *"Ibid."*, **24**, p. 3681 (1991).
20. Siqueira D. F., Pitsikalis M., Hadjichristidis N., Stamm M., *"Langmuir"*, **12**, p. 1631 (1996).
21. Siqueira D. F., Breiner U., Stadler R., Stamm M., *"Ibid."*, **11**, p. 1680 (1995).
22. Shybanova O., Voronov S., Bednarska O. et al., *"Macromol. Symp."*, **164**, p. 211 (2001).
23. Shybanova O. B., Medvedevskikh Yu. G., Voronov S. A., *"Vysokomolekulyarnyye soyedinieniya"*, **43** A (11), p. 1964 (2001) *(in Russian)*.
24. Medvedevskikh Yu. G., *"Condensed Matter Physics"*, **4** (2) (26), p. 209 (2001).
25. Medvedevskikh Yu. G., *"J. Appl. Pol. Sci."*, **109** (4), p. 2472 (2008).
26. Medvedevskikh Yu. G., In.: *"Conformation of Macromolecules"* / *Ed. by Yu. G. Medvedevskikh, S. A. Voronov, G. E. Zaikov, New York: Nova Science Publishers*, p. 35 (2007).

CHAPTER 5

PHENOMENOLOGICAL COEFFICIENTS OF THE VISCOSITY FOR LOW-MOLECULAR ELEMENTARY LIQUIDS AND SOLUTIONS

YU. G. MEDVEDEVSKIKH and O. YU. KHAVUNKO

CONTENTS

SUMMARY

Starting from the general phenomenological determinations of the sub-stance flow under the action of chemical potential gradient and from the analysis of shearing forces and corresponding strains appearing into the flow it was determined that the viscosity coefficient of the pure liquid is ordered to the ratio $\eta = 3RT\tau/V$ and the expression $\eta_i = 3RT\tau_i/V_i$ is correct for the component of the solution. The preexponential factor τ_0 under ex-pression of the characteristic time t of the viscous flow is determined not only by the frequency of the fluctuating motion of the particles in quasi-lattice but also by the entropic factor. Obtained expression for the activa-tion entropy $\Delta S^* = \Delta H^*/T^*$ explains the low values $\tau_0 \ll 2h/kT$ for the associated liquids and the antibate relationship between τ_0 and the acti-vation energy for the viscous flow. It was proposed the expressions per-mitting to calculate the coefficients of the self-diffusion and the diffusion upon corresponding coefficients of the viscosity for pure liquids and solu-tions. An analysis of the Maxwell's equation and also of the deformation rates of the conformational volume of polymeric chains and their rotation permitted to mark out the frictional and elastic coefficients of the viscosity of high molecular one component liquid. It was shown, that exactly elastic coefficient of the viscosity is the gradient dependent value.

5.1   INTRODUCTION

The temperature dependence of the viscosity coefficient for the low-molecular elementary liquids is good described by the *Arrhenius'* empirical equation:

$$\tau = \tau_0 \exp\{E/RT\} \tag{1}$$

where: E is the activation energy of the viscous flow. However, the theo-retical interpretation of the preexponential factor A is struck against the appreciable difficulties which were not overcame to the present time.

Among the molecular-kinetic theories of low-molecular liquids sub-mitting to the *Newton* equation, let us mark out two main formulated by *Frenkel* [1–2] and by *Eyring* [3–5].

In accordance with the *Frenkel's* theory under the presence of a shear strain σ the rapid layer of liquid entrains the slow one with the velocity rate $\Delta\vartheta$:

$$\Delta = \vartheta\sigma\delta^2 \tag{2}$$

Here u is the mobility of particle and δ is the interparticle distance.

Substituting of $\Delta\vartheta$ on $\frac{d\vartheta}{dy}\delta$, where $d\vartheta/dy$ is the velocity gradient of the hydrodynamic flow along y axis, which is normal to the flow, and using the *Newton's* equation in a form $\eta\frac{d\vartheta}{dy} = \sigma$, it was obtained the following ration by *Frenkel*:

$$\eta = (u\delta)^{-1} \tag{3}$$

The substitution of the *Einstein's* equations into Eq. (3)

$$u = D / kT, \tag{4}$$

$$D = \delta^2 / 6\tau \tag{5}$$

leads to the final result

$$\eta = 6kT\tau / \delta^3 \tag{6}$$

Here D is the coefficient of the diffusion; t is the characteristic time of the particles transport from the one equilibrium state into other, which can be also called as the characteristic time of the viscous flow.

Accordingly to *Frenkel* t is determined by the oscillation frequency into the quasi-lattice of liquid and by the probability of the whole formation, i.e., corresponding free volume needed for the particle transport:

$$\tau = \tau_0 \exp\{E / RT\} \tag{7}$$

Thereby, here τ_0 is the characteristic time of the oscillating movement of a particle into the quasi-lattice of liquid; E is an energy of the whole formation or of the free volume into liquid, which is necessary in order to pass the particle from the one position of equilibrium into another one.

After the integration of Eqs. (6) and (7) the *Frenkel's* equation becomes as follows

$$\eta = \frac{6kT}{\delta^3} \tau_0 \exp\{E / RT\} \qquad (8)$$

Under the calculations it was assumed by *Frenkel* $\tau_0 = 10^{-13} s$ that is equal to the determination

$$\tau_0 = h / kT. \qquad (9)$$

However, the calculations of the preexponential multiplier in Eq. (8) sometimes give the values exceeding the experimental ones on 2–3 orders. These divergences were explained by *Frenkel* with dependence of E on temperature; however, such phenomenon is observed utmost seldom.

In molecular-kinetic theory of *Eyring* the starting principle consists in fact, that the action of force causing the liquid flow, decreases the height of the energy barrier at the movement of particle into forward direction and increases it into back direction. The rate constant k of the particle's transfer via the potential barrier is described by the standard equation of the theory of absolute rates of chemical reactions

$$\kappa = \frac{kT}{h} \frac{Q^*}{Q} \exp\{-E / RT\} \qquad (10)$$

where Q and Q^* are the statistical sums of the particle per unit of volume in main and activated states respectively.

Under great following simplifications the force of shift disappears from the equation of transfer. As a result, the expression for the viscosity coefficient accordingly to *Eyring* becomes as follows

$$\eta = \frac{hN_A}{V} \frac{Q}{Q^*} \exp\{E / RT\} \qquad (11)$$

where: V is the molar volume of liquid.

Assuming, that the statistical sum Q^* of particle into the activated state devoid of one freedom degree of the transitional movement, *Eyring* writes:

$$\frac{Q}{Q^*} = \frac{(2\pi mkT)^{1/2}}{h} v_f^{1/3} \tag{12}$$

where v_f is free volume per one particle.

Let's note, however, that in accordance with the starting Eq. (10) Q^* is already devoid of one freedom degree of the oscillating movement, otherwise the co-multiplier kT/h would be not appeared in Eq. (10). To devoid of once more freedom degree, *namely*, transitional one is physically absolutely unjustified.

Although the Eq. (12) in the presented case is the fitting of theory under the result, let us use it for the determination of characteristic time of the translational movement

$$\tau_0 = \left(\frac{2\pi m}{kT}\right)^{1/2} v_f^{1/3} \tag{13}$$

By combining of Eqs. (12) and (13) into (11) it can be obtained

$$\eta = \frac{kN_A T}{V} \tau_0 \exp\{E/RT\} \tag{14}$$

In accordance with the *Frenkel's* Eq. (7) such expression can be written as follows:

$$\eta = RT\tau/V \tag{15}$$

If to assume δ^3 as the volume falling per one particle, we will obtain from the Eq. (6) the equation analogous to Eq. (15) one:

$$\eta = 6RT\tau/V \tag{16}$$

Thus, in spite of the difference in molecular-kinetic approaches of *Frenkel* and *Eyring* to the analysis of the viscosity coefficient for simple liquids, they lead to the ratios containing the same phenomenological value $RT\tau/V$. The difference between Eqs. (15) and (16) is visualized in numerical coefficient and physical interpretation of τ_0 determined by the Eqs. (9) and (13). This difference is quite substantially, since τ_0 of the translational movement is on two orders higher than the τ_0 of the oscillating movement. At the analysis of the solution viscosity *Eyring* and *Frenkel* do not use the notion "*a coefficient of viscosity of solution component*", preferring to describe the viscosity of solution via the coefficients of viscosity for pure liquids. At this, it is assumed that the viscosity of solution can be described by the equations analogous to Eqs. (8) and (11), but, at the same time, a free activation energy (accordingly to *Eyring*) or an activation energy (accordingly to *Frenkel*) is declared as the function of the solution composition. This leads of *Eyring* to the equation:

$$\ln \eta = N_1 \ln \eta_1 + N_2 \ln \eta_2 \qquad (17)$$

which was proposed by *Arrhenius* and *Kendall* earlier [6].

It was proposed more complicated equation for the binary solution by *Frenkel*

$$\ln \eta = \frac{1}{2} N_1^2 \ln \eta_1 + \frac{1}{2} N_2^2 \ln \eta_2 + N_1 N_2 \ln \eta_{12}. \qquad (18)$$

In these equations N_i is molar part of the component; η_i is its coefficient of viscosity in pure liquid state; η_{12} is additional viscosity, which, in accordance with *Frenkel*, reflects the difference in energies of interaction of particles by the first and the second kinds between themselves and between the particles of the same kind.

In conclusion of the presented short review let us note, that the molecular-kinetic analysis of *Frenkel* and *Eyring* as to viscosity coefficients for pure liquids leads to the Eqs. (15) and (16) containing the general phenomenological factor $RT\tau/V$, which requires the substantiation from more general considerations. Coefficients of viscosity of the solution components in this approach don't reveal. The characteristic time of the viscous flow is remained undefined. We will be solving of these problems in the following chapters.

5.2 COEFFICIENT OF VISCOSITY FOR LOW-MOLECULAR PURE LIQUID

Phenomenological determination of the stationary flow J of substance, the moving force of which is the gradient of the chemical potential μ [7, 8], let us accept ad the starting position for the analysis. Let us write of these flows along the direction of the x and y axis:

$$J_x = -\frac{L}{RT}c\frac{\partial \mu}{\partial x}, \quad J_y = -\frac{L}{RT}c\frac{\partial \mu}{\partial y} \tag{19}$$

where L is the transfer coefficient having the dimension of the diffusion coefficient; c is the molar-volumetric concentration of particles.

For pure liquid at T = *const* the chemical potential μ is the function only on pressure P. That is why it can be written as $n\mu/nx = VnP/nx$, where V is a molar volume. Since $Vc = 1$, instead of the Eq. (19) we will writing

$$J_x = -\frac{L}{RT}\frac{\partial P}{\partial x} \qquad J_y = -\frac{L}{RT}\frac{\partial P}{\partial y} \tag{20}$$

The flows can be expressed also via the transfer rate u and concentration c: $J_i = cu_i$. Then, with taking into account of the Eq. (20) we have

$$u_x = -\frac{LV}{RT}\frac{\partial P}{\partial x}, \quad u_y = -\frac{LV}{RT}\frac{\partial P}{\partial y} \tag{21}$$

By differentiating of u_x upon y, and u_y upon x, we will obtain

$$\frac{\partial u_x}{\partial y} + \frac{\partial u_y}{\partial x} = -2\frac{LV}{RT}\frac{\partial^2 P}{\partial x \partial y} \tag{22}$$

If η is the coefficient of the liquid viscosity, then accordingly to the *Newton's* equation

$$\eta\left(\frac{\partial u_x}{\partial y} + \frac{\partial u_y}{\partial x}\right) = \sigma_{xy} = \sigma_{yx} \tag{23}$$

where $\sigma_{xy} = \sigma_{yx}$ are the shift components of the stress tensor; the first index points out direction of the component of force, and the second one points out direction of the normal to the plate of the application of force.

It's follows from the comparison of Eqs. (22) and (23):

$$-2\eta \frac{LV}{RT} \frac{\partial^2 P}{\partial x \partial y} = \sigma_{xy} = \sigma_{yx} \tag{24}$$

Next, in liquid let us separate the elementary cube with the edges δ. The sectional elevation of this cube by the plate xy is shown on Fig. 1a.

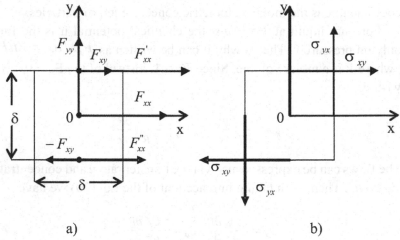

a)   b)

FIGURE 1   Scheme of the formation of shear forces F_{xy} and tensions σ_{xy} under dependence of normal force F_{xx} on coordinate y.

Let in center of cube the gradient of pressure in direction x is equal to $-\partial P / \partial x$.

Since $-\partial P / \partial x$ is the force acting on unit of the volume of liquid, full force F_{xx}, acting on the whole volume δ^3 of cube in direction of the x *axe* will be equal to

$$F_{xx}(x = 0, y = 0) = -\frac{\partial P}{\partial x} \delta^3 \tag{25}$$

Tangential or shear forces are appeared as a result of the dependence of F_{xx} on y. Let write the new forces F_{xx} at $x = 0$ as $y = \frac{\delta}{2}$ and $y = -\frac{\delta}{2}$:

$$F'_{xx}\left(x=0, y=\frac{\delta}{2}\right) = -\frac{\partial P}{\partial x}\delta^3 - \frac{\partial^2 P}{\partial x \partial y}\delta^4 / 2, \tag{26}$$

$$F''_{xx}\left(x=0, y=-\frac{\delta}{2}\right) = -\frac{\partial P}{\partial x}\delta^3 + \frac{\partial^2 P}{\partial x \partial y}\delta^4 / 2. \tag{27}$$

The difference in these forces creates the shear forces F_{xy}, applied to the top and the bottom of cube faces, normal to the axe y (*see* Fig. 1a):

$$F_{xy}\left(x=0, y=\frac{\delta}{2}\right) = F'_{xx} - F''_{xx} = -\frac{\partial^2 P}{\partial x \partial y}\delta^4, \tag{28}$$

$$F_{xy}\left(x=0, y=-\frac{\delta}{2}\right) = F''_{xx} - F'_{xx} = \frac{\partial^2 P}{\partial x \partial y}\delta^4. \tag{29}$$

These forces create the shear tensions $\sigma_{xy} = F_{xy} / \delta^2$: on the top face

$$\sigma_{xy} = -\frac{\partial^2 P}{\partial x \partial y}\delta^2 \tag{30}$$

and on the bottom face

$$\sigma_{xy} = \frac{\partial^2 P}{\partial x \partial y}\delta^2 \tag{31}$$

It is follows from this, that under the chosen direction of force F_{xx}

$$-\frac{\partial^2 P}{\partial x \partial y} = \frac{\sigma_{xy}}{\delta^2} \tag{32}$$

The stress tensor is symmetrical, that can be easy demonstrated by the analogous analysis of force F_{yy} depending on x, that is why $\sigma_{xy} = \sigma_{yx}$, as it is shown on Fig. 1b.

By substituting of Eq. (32) in (24), we will obtain

$$\eta = \frac{1}{2}\frac{RT}{LV}\delta^2 \tag{33}$$

Next using the *Einstein's* ratio for the transfer coefficient

$$L = \delta^2 / 6\tau \tag{34}$$

which takes into account the randomness of the particle wandering upon the all three directions of space, we finally find:

$$\eta = 3RT\tau / V \tag{35}$$

5.3 PARTIAL COEFFICIENTS OF VISCOSITY OF THE LOW-MOLECULAR SOLUTION COMPONENTS

The flows of i-component of the solution along x and y axis let determine accordingly to the Eq. (19):

$$J_{ix} = -\frac{L_i}{RT} c_i \frac{\partial \mu_i}{\partial x}, \quad J_{iy} = -\frac{L_i}{RT} c_i \frac{\partial \mu_i}{\partial y} \tag{36}$$

Due to the difference in the transfer coefficients L_i of the solution's components it can be formally supposed that into the hydrodynamic flow it can be appeared the conditions like to the components separation in the baromembrane process [8]. In the last case these conditions are realized at the expense of a high gradient pressure upon membrane's layer and a great difference in transfer coefficients L_i and L_j via the membrane.

We, however, will be suppose, that at the viscosity change the hydrodynamic flows are enough weak and the effect of the components separation can be neglected. Consequently, a liquid solution is homogeneous upon composition and the all derivatives by dc_i/dx type are equal. Then, although the chemical potential μ_i is the function on the solution composition via the thermodynamic activity of component, the derivatives $\partial \mu_i / \partial x$ are the function only on the gradient of pressure. So, at $T = const$ and presented composition of solution, $\partial \mu_i / \partial x = \overline{V_i} \partial P / \partial x$, where $\overline{V_i}$ is the partial-molar volume of i component of solution. Since

$$c_i \overline{V_i} = \phi_i \tag{37}$$

where φ_i is the volumetric part of presented component, the Eq. (36) can be written as

$$J_{ix} = -\frac{L_i}{RT}\phi_i\frac{\partial P}{\partial x} \quad J_{iy} = -\frac{L_i}{RT}\phi_i\frac{\partial P}{\partial y}. \tag{38}$$

Stating of these flows in a form $J_{ix} = c_i u_{ix}$, $J_{iy} = c_i u_{iy}$, and taking into account Eqs. (37) and (38) it can be obtained:

$$\frac{\partial u_{ix}}{\partial y} + \frac{\partial u_{iy}}{\partial x} = -2\frac{L_i\bar{V}_i}{RT}\frac{\partial^2 P}{\partial x\partial y} \tag{39}$$

Since nP/nx is a force applied to the unit of the volume of solution, then $\phi_i\partial P/\partial x$ is a force, applied to the presented component into unit of the volume. Thereby, it can be supposed, that the endowment of the presented component into general viscosity of solution η depends not only on the partial coefficient of viscosity η_i but also on volumetric part φ_i. That is why we postulate the dependence:

$$\eta = \sum_i \eta_i\phi_i \tag{40}$$

Thus, the endowment of i component into the viscosity of solution is determined by the value $\eta_i\varphi_i$, that is why, the *Newton's* equation for each component of the solution is necessary to write as

$$\eta_i\phi_i\left(\frac{\partial u_{ix}}{\partial y} + \frac{\partial u_{iy}}{\partial x}\right) = \sigma_{ixy} = \sigma_{iyx} \tag{41}$$

where $\sigma_{ixy} = \sigma_{iyx}$ is the endowment of i component into corresponding full shear stresses $\sigma_{xy} = \sigma_{yx}$, like this

$$\sigma_{xy} = \sum_i \sigma_{ixy} \tag{42}$$

It is following from the comparison of Eqs. (39) and (41)

$$2\eta_i\phi_i \frac{L_i\bar{V_i}}{RT}\frac{\partial^2 P}{\partial x\partial y} = \sigma_{ixy} = \sigma_{iyx} \tag{43}$$

Let us again refer to the elementary cube by volume δ^3 (Fig. 1). In the center of this cube the force into direction of x axe acts on the presented component; this force, as it was noticed, is equal to $-\varphi_i\partial P/\partial x$. Taking into account the volume of cube, we have

$$F_{ixx}(x=0,y=0) = -\phi_i\frac{\partial P}{\partial x}\delta^3 . \tag{44}$$

The force F_{ixx} depends on coordinate y, that is why we write the new forces at x = 0, but at y = $\delta/2$ and y = $-\delta/2$

$$F_{ixx}(x=0,y=\frac{\delta}{2}) = -\phi_i\frac{\partial P}{\partial x}\delta^3 - \phi_i\frac{\partial^2 P}{\partial x\partial y}\delta^4/2 \tag{45}$$

$$F_{ixx}(x=0,y=-\frac{\delta}{2}) = -\phi_i\frac{\partial P}{\partial x} + \phi_i\frac{\partial^2 P}{\partial x\partial y}\delta^4/2 \tag{46}$$

The difference in these forces creates the shearing forces F_{ixy}, applied to the cube faces, normal to y axis, which determine the shear stresses $\sigma_{ixy} = F_{ixy}/\delta^2$. In particular, into direction of F_{ixx} the shear stress on the top face of cube will be equal to:

$$\sigma_{ixy} = -\phi_i\frac{\partial^2 P}{\partial x\partial y}\delta^2 \tag{47}$$

Using of this ratio in Eq. (43), we will obtain:

$$\eta_i = \frac{1}{2}\frac{RT}{L_i\bar{V_i}}\delta^2 . \tag{48}$$

Expressing the transfer coefficient L_i via corresponding *Einstein's* ratio $L_i = \delta^2/6\tau_i$ we will finally obtained:

$$\eta_i = 3RT\tau_i/\bar{V_i} \tag{49}$$

Thus, postulated dependence (Eq. 40) leads to the Eq. (49) for the coefficient of the viscosity of solution in the same form as for the pure liquid. However, the viscosity coefficient η_i is the function on a composition of the solution.

The substitution of Eq. (49) into (40) with taking into account of Eq. (37) permits to express the viscosity of solution by more convenient from the practical point of view ratio

$$\eta = 3RT \sum_i c_i \tau_i \qquad (50)$$

For binary solution it can be written:

$$\eta = 3RT(c_1 + c_2)(N_1\tau_1 + N_2\tau_2) \qquad (51)$$

where N_i is molar part of the component.

Since $c_1 + c_2 = V^{-1}$, where V is a molar volume of the solution, equal to $(M_1N_1 + M_2N_2)/\rho$, it's follows from the Eq. (51) that

$$\eta = 3RT\left(\frac{\rho}{M_1N_1 + M_2N_2}\right)(\tau_1 + N_2(\tau_2 - \tau_1)) \qquad (52)$$

where M_i are molar masses of components, ρ is a density of solution at presented composition.

An Eq. (52) permits to find the numerical values of τ_i and their dependence on composition of the solution based on the experimental values of viscosity η.

5.4 CHARACTERISTIC TIME OF THE VISCOUS FLOW

As it can be seen from the phenomenological Eqs. (35) and (49), the main problem at the viscosity coefficient calculation is the determination of the characteristic time t of the viscous flow. In accordance with the *Frenkel t* can be described by the Eq. (7), in which τ_0 is determined either by vibration freedom degree of liquid's particles or by the translational one, in others words by the Eqs. (9) or (13), respectively.

In order to compare the experimental values of τ_0 with the calculated ones accordingly to Eqs. (9) and (13), we described the temperature dependence of viscosity for a series of the *n*-alkanes and *n*-alcohols, and also for water on the basis of referenced data via the *Arrhenius* equation in form:

$$\ln\eta = \ln A + E/RT \tag{53}$$

Coefficients of this equation are represented in Table 1. On the basis of viscosity values at temperature 293 K, it were calculated accordingly to Eq. (35) the values τ_{293}, and after that accordingly to Eq. (7) the values τ_0. The results of calculations are represented in Table 1.

At T = 293 K the characteristic time of the oscillating movement is equal to $h/kT = 1.63 \cdot 10^{-13}$ s, and of the translational motion $(2\pi m/kT)^{1/2} v_f^{1/3}$ is approximately on two orders more.

TABLE 1   Referenced and calculated data concerning to the viscosity of some liquids.

Liquid	M, g/ mole	ρ, 10^6 g/ m^3	ΔH_{ev}, kJ/ mole	η_{293K}, $10^{-3}Pa \cdot s$	$-lnA$ (Pa·c)	$\frac{E}{R}$, K	τ_{293K}, 10^{-12} s	τ_0, 10^{-13} s	$\frac{\Delta S^\circ}{R}$	T, K	$D \cdot 10^9$, M^2/s
Pentane C_5H_{12}	72	0.626	26.43	0.229	11.29	851	3.61	1.98	0.50	—	15.4
Hexane C_6H_{14}	86	0.655	31.55	0.320	11.12	900	5.75	2.67	0.20	—	10.6
Heptane C_7H_{16}	100	0.684	36.55	0.409	11.10	970	8.21	3.04	0.07	—	8.0
Octane C_8H_{18}	114	0.702	41.48	0.540	11.19	1070	12.03	3.10	0.05	—	5.8
Nonane C_9H_{19}	128	0.718	39.92	0.710	11.32	1195	17.37	2.94	0.11	—	4.3
Methanol CH_3OH	32	0.793	38.45	0.584	11.83	1285	3.23	0.40	2.10	612	8.6
Ethanol C_2H_5OH	46	0.789	42.01	1.190	12.63	1725	9.52	0.26	2.53	682	3.7
Propanol C_3H_7OH	60	0.804	48.12	2.256	13.38	2135	23.10	0.16	3.02	707	1.8

Buthanol C_4H_9OH	74	0.810	52.3	2.950	13.74	2315	37.00	0.14	3.15	735	1.3
Pentanol $C_5H_{11}OH$	88	0.814	56.94	4.140	15.88	3045	61.42	0.019	5.15	591	0.9
Water H_2O	18	0.997	40.66	1.005	13.22	1865	2.49	0.043	4.33	431	6.5

That is why, the experimental values $\tau_0 \ll (2\pi m/kT)^{1/2} v_f^{1/3}$. At the same time, for the n-alkanes τ_0 is some more h/kT, although is by the same order, but for the associated liquids, n-alcohols and water, τ_0 is considerable lesser h/kT. Besides, in a series of the n-alcohols it is observed a well-defined slope opposition of the dependence between the activation energy of the viscous flow and τ_0: the more is E, the lesser is τ_0. Both factors, that is, $\tau_0 \ll h/kT$ and observed slope opposition compensation effect, cannot be explained via the approximations of *Eyring* and *Frenkel*.

In the *Eyring's* theory of the absolute reactions rates there are three essential lacks: a) the concentration of the activated complexes can be found from the consideration of condition of their equilibrium with the initial (or final) substances; b) the activated complex devoided of one freedom degree along the coordinate of the reaction; c) the transmission coefficient is the empirical co-multiplier.

The approach keeping the main advantages of the theory of absolute reactions rates but eliminating the listed above lacks is described in detail in Ref. [9]. Here let us limit only by its short description with the specific aim of analysis of characteristic time of viscous flow.

Let the elementary reversible reaction

$$A \underset{\underline{\kappa}}{\overset{\overline{\kappa}}{\rightleftarrows}} B \qquad (54)$$

proceeds via the activated complex C, general for initial and final substances. Then it can be written as:

$$A \underset{v_1}{\overset{\kappa_1}{\rightleftarrows}} C \underset{\kappa_1}{\overset{v_2}{\rightleftarrows}} B \qquad (55)$$

In accordance with Eq. (55), the activated complex C has a right to roll back from the top of the potential barrier into the both potential holes with the frequencies v_1 and v_2, which aren't the activated parameters. Parameters κ_1 and κ_2 are activate, in other words, they depend on the value of the potential barrier, and determine the frequencies of the activated complexes formation from the initial and final substances, respectively.

Let a, *b* and c are the concentrations A, B and C. The rate of the elementary reaction (Eq. 54) accordingly to the law of mass action:

$$\upsilon = \vec{\kappa}a - \overleftarrow{\kappa}b \tag{56}$$

Is strictly defined, in others words has the physical sense, only under condition $dc / dt = 0$, to which the stationary concentration of activated complexes c_s corresponds:

$$c_s = (\kappa_1 a + \kappa_2 b) / (v_1 + v_2). \tag{57}$$

In this case, in accordance with the scheme (Eq. 55),

$$\upsilon = \kappa_1 a - v_1 c_s = v_2 c_s - \kappa_2 b \tag{58}$$

By substituting of Eq. (57) into Eq. (58), we will obtain

$$\upsilon = \frac{\kappa_1 v_1}{v_1 + v_2} a - \frac{\kappa_2 v_1}{v_1 + v_2} b \tag{59}$$

If into the system an equilibrium *Maxwell–Boltzmann* distribution of the energy between the particles and their freedom degrees is kept, then in accordance with the law of mass action the following ratios should be performed:

$$\kappa_1 / v_1 = K_1, \quad \kappa_2 / v_2 = K_2 \tag{60}$$

where K_1 and K_2 are equilibrium constants of the activated complexes formation from the initial and final substances, respectively.

By substituting of these ratios in Eq. (59) and comparing them with Eq. (56), we will obtain for the constants rates of direct and reverse elementary reaction (Eq. 54) the following expressions:

$$\vec{\kappa} = \frac{v_1 v_2}{v_1 + v_2} K_1, \quad \overleftarrow{\kappa} = \frac{v_1 v_2}{v_1 + v_2} K_2. \tag{61}$$

As to the analysis of the characteristic time of viscous flow the constant rate κ of the particles transfer via the potential barrier can be presented by analogous (Eq. 61) expression:

$$\kappa = \frac{v_1 v_2}{v_1 + v_2} K^* \tag{62}$$

where K^* is the constant of equilibrium of activated complex with particles into the main state. It can be described by usual thermodynamical ratio via the standard entropy ΔS^* and enthalpy ΔH^* of activation:

$$K^* = \exp\{\Delta S^* / R\} \exp\{-\Delta H^* / RT\}, \tag{63}$$

Let the seaward run of the activated complex into the both directions of the transfer is determined by the oscillating freedom degree in such a way that $v_1 = v_2 = kT / h$. Then:

$$\kappa = \frac{1}{2} \frac{kT}{h} \exp\{\Delta S^* / R\} \exp\{-\Delta H^* / RT\} \tag{64}$$

Here well defined the value of the transmission coefficient. Characteristic time of the transfer is inversely proportional to κ that is why

$$\tau = 2 \frac{h}{kT} \exp\{-\Delta S^* / R\} \exp\{\Delta H^* / RT\}, \tag{65}$$

and for τ_0 we have the following expression:

$$\tau_0 = 2 \frac{h}{kT} \exp\{-\Delta S^* / R\} \tag{66}$$

In accordance with the data of Table 1, the experimental values of τ_0 for n-alkanes are practically agreed with the value $2h/kT = 3.27 \cdot 10^{-13}$ s.

For the associated liquids $\tau_0 \ll 2h/kT$ that points on essentially more value of positive entropy of activation ΔS^*. These values $\Delta S^*/R$, calculated accordingly to Eq. (66), represented in Table 1. For the homologous series of the n-alcohols is accurate observed the symbate dependence of ΔS^* on E. For the n-alkanes such relationship isn't observed, since the values of ΔS^* for them are little and remain via the ranges of the inaccuracy of their determination based on experimental values τ_0.

In accordance with the conclusion (Eq. 62) the activated complex has not only the same number, but also the same freedom degree as the particles into the main state. That is why in the presented case it cannot be obtain the $\Delta S^* > 0$ at the expense of the freedom degrees redistribution at the transfer of particle from the main state into the activated one. Since at given P and T the activation is accompanied by the energy increasing, also the entropy should be increased like to the fact that the entropy is increased at the phases transition by the first kind at $\Delta H > 0$.

Thus, we suppose, that via the activation process at viscous flow of liquid the change of ΔS^* can be expressed via ΔH^* as same as at the phase transition by the first kind. Activation process at given P and T is nonequilibrium, however theoretically it can be always found at given P such hypothetic temperature T*, at which the activation transition will be equilibrium. Therefore, at this temperature T*

$$\Delta S^* = \Delta H^* / T^* \tag{67}$$

Then the Eq. (66) can be rewritten as:

$$\tau_0 = 2\frac{h}{kT}\exp\left\{-\frac{\Delta H^*}{RT^*}\right\}. \tag{68}$$

Calculated values of T* accordingly to Eq. (67) for the n-alcohols and water are represented in Table. For homological series of n-alcohols the values of T^* are not strongly differed one from other, and exactly this explains the experimentally observed the slope opposition dependence of τ_0 on $\Delta H^* \approx E$.

In conclusion of the presented chapter it is necessary to note, that the activation entropy by Eq. (67) type occurs also in chemical processes as one among terms of the general activation entropy. This fact is proved both by rapidly observed compensation effects in a range of the single-type chemical reactions, and by the presence of so-called quick reactions, description of which is not kept within the *Eyring's* theory.

5.5   CALCULATION OF THE DIFFUSION COEFFICIENTS BASED ON THE COEFFICIENTS OF VISCOUSITY

Experimental determination of the diffusion coefficients D is enough complicated and labor intensive process, whereas the experimental determination of the coefficients of viscosity doesn't strike the great complications.

Established long ago the empirical *Walden's* rule

$$\eta D \approx const \tag{69}$$

doesn't give the estimation of the value *const* and is approximate, but points on the possibility of D calculation based on the experimental values η.

For pure liquids the relationship between the transfer coefficient L or coefficient of self-diffusion D is determined via the ratio (Eq. 33), which can be rewritten in a form of the *Walden's* rule $(L=D)$

$$\eta D = \frac{1}{2}\frac{RT}{V}\delta^2 \tag{70}$$

Expressing the transfer area δ^2 via the molar volume of liquid

$$\delta^2 = (V / N_A)^{2/3}, \tag{71}$$

for the coefficient of the self-diffusion we will obtain the expression:

$$D = \frac{1}{2} \frac{RT}{\eta N_A^{2/3}} \left(\frac{\rho}{M} \right)^{1/3}$$

(72)

The calculated values of D accordingly to Eq. (72) with the use of experimental data of η at T = 293 K are represented in Table 1. As we can see, the calculated data of D have a typical order for the liquids by similar kind, well defined reflect the dependence of D on molar mass in presented homolytical range and on nature of homolytical range.

For the solutions the determination of coefficients of diffusion of components upon the viscosity is more complicated task. In accordance with Eqs. (48) and (71) we have

$$D_i = \frac{1}{2} \frac{RT}{\eta_i \overline{V_i}} \left(\frac{V}{N_A} \right)^{1/3}$$

(73)

Here V is a molar volume of the solution, so $V = M / \rho$, $M = \sum_i M_i N_i$ is a molar mass of the solution.

As we can see, the calculation of the coefficient of diffusion of the solution's component accordingly to Eq. (73) needs the knowledge, first of all, the coefficient of the viscosity of the presented component and its partial-molar volume.

5.6 FRICTIONAL AND ELASTIC COEFFICIENTS OF VISCOUSITY OF HIGHMOLECULAR LIQUIDS

In contrast to the low-molecular liquids the high-molecular ones, *namely* polymeric solutions and melts, are called by "viscoelastic". This means that the measured or effective viscosity of the polymeric solutions should has *the frictional component*, caused by the forces of friction between the layers of liquid, which are moving under the action of gradient rate of the hydrodynamic flow with different rates, and *the elastic component*, caused by the property of conformation of polymeric chain to do the resistance to the shear strain. Such fact, that a so-called anomalous behavior of the polymeric solutions, that is the dependence of their effective viscosity on gradient rate of the hydrodynamic flow is not properly for the low-molecular liquids, permits to assume, that exactly the elastic component of the viscosity of polymeric solutions is gradiently depended value.

Let consider of this problem for the single-component high-molecular liquid, for example for polymeric melt [10].

The *Newton's* equation in a form (Eq. 23) in this case will be determine the shear stress only via the frictional coefficient of viscosity η_f:

$$\sigma_{xy} = \sigma_{yx} = \eta_f \left(\frac{\partial u_x}{\partial y} + \frac{\partial u_y}{\partial x} \right), \tag{74}$$

With taking into account of the elastic properties of conformation of polymeric chains the shear stress should be described by the *Maxwell's* equation

$$\sigma_{xy} = \sigma_{yx} = \eta_f \left(\frac{\partial u_x}{\partial y} + \frac{\partial u_y}{\partial x} \right) + \mu \theta_{xy} \tag{75}$$

Here: θ_{xy} is an angle of shear in the plane xy, μ is the shear module, determined in Refs. [11, 12] for the diluted and concentrated polymeric solutions and melts.

Let input the gradient rate of the hydrodynamic flow

$$g = \frac{\partial u_x}{\partial y} + \frac{\partial u_y}{\partial x}. \tag{76}$$

and let rewrite the Eq. (75) in more simple form

$$\sigma = \eta_f g + \mu \theta \tag{77}$$

Thus, the shear stress calculated during the experiment, is determined both by the shear of liquid's layers one relatively other and by the deformation shear of the conformational volumes of polymeric chains. However, the measured shear stress is correlated with the known gradient of rate of the hydrodynamic flow in a form of the *Newton's* Eq. (74), but not in a form of the *Maxwell's* Eq. (75):

$$\sigma = \eta g \cdot \tag{78}$$

Therefore, in Eq. (78) η is the measured effective viscosity. Comparing the Eqs. (77) and (78), for the effective viscosity we will obtained the equation

$$\eta = \eta_f + \mu\theta/g \tag{79}$$

Usually the *Maxwell's* model is represented as sequentially combined the elastic spring and damper, corresponding respectively for the second and for the first terms in Eq. (79). As to the polymeric solutions this model has two essential lacks. First is the fact that it supposes that the spring instanter responses on the external action, whereas the polymeric chain has enough more relaxation time and its response on the external action is extended via time. Secondly, in contrast to the spring into the *Maxwell's* model, the polymeric chain into the solution under the action of the gradient of rate of the hydrodynamic flow is not only deformated but also is rotated. This leads to fact that the shear forces act for the all time on the different sections of the conformational volume. Therefore, the value of shear deformation of the conformational volume of polymeric chain will be depend not only on the applied shear forces, but also on the rotation rate of the polymeric ball. Taking into account of the above-said, the gradient rate of the hydrodynamic flow g let input as two parts, first of which determines the rate of the deformation shear g_s of the conformational volume of polymeric chain, and the second determines an angle rate of its rotation g_r at which the frozen equilibrium conformational state of the polymeric chain is retained, as it is formulated for example in the model of the hard bended *Kuhn's* wire [13]:

$$g = g_s + g_r . \tag{80}$$

Let $\theta = \theta_s$ is a limiting or stationary value of the shear angle corresponding to the presented external action g_s at infinite time of action. Then the rate of shear can be described by simple kinetic equation

$$g_s = \frac{d\theta}{dt} = k_s(\theta_s - \theta) . \tag{81}$$

The value inverse to the constant of the shear rate is the characteristic time of shear: $t^* = k_s^{-1}$. That is why let rewrite the Eq. (81) as

$$\frac{d\theta}{dt} = \frac{1}{t^*}(\theta_s - \theta) \tag{82}$$

Integrating the Eq. (82) via the limits from $\theta = 0$ at $t = 0$ till the value θ, achieved at $t = t_v$, where t_v is the characteristic time of the external action, we will obtained

$$\theta = \theta_s\left(1 - \exp\left\{-\frac{t_v}{t^*}\right\}\right) \tag{83}$$

It follows from this, that $\theta \to \theta_s$ at $t_v \to \infty$, since t^* is the property of the polymeric chain does not depending on time of the external action.

Combining the Eqs. (82) and (83), we will obtained the expression for the rate of shear deformation:

$$g_s = \frac{\theta_s}{t^*}\exp\left\{-\frac{t_v}{t^*}\right\} \tag{84}$$

Let suppose [10] that the rotation and shear deformation of the conformational volume of polymeric chains inherently have the same reparation mechanism [14], which is realized via the segmental movement. In such a case the angular rotation rate has the same characteristic time t^*, but at this it does not depend on the current rotation angle. That is why it can be accepted as equal to the initial rate of shear at $t = 0$ and respectively $\theta = 0$. This means accordingly to Eq. (84), that

$$g_r = \frac{\theta_s}{t^*}. \tag{85}$$

Combining the Eqs. (84) and (85) into (80), we will obtain

$$g = \frac{\theta_s}{t^*}\left(1 + \exp\left\{-\frac{t_v}{t^*}\right\}\right) \tag{86}$$

Next, by substituting the Eqs. (83) and (86) in (79), we will found

$$\eta = \eta_f + \eta_e^0 \frac{1 - \exp\left\{-t_v/t^*\right\}}{1 + \exp\left\{-t_v/t^*\right\}}. \tag{87}$$

Here

$$\eta_e^0 = \mu t^* \tag{88}$$

depends only on the properties of the macromolecule that is why it represents by itself the coefficient of the elastic component of viscosity, the endowment of which into the effective viscosity, however depends on the gradient rate of the hydrodynamic flow. The value

$$\eta_e = \eta_e^0 \frac{1 - \exp\left\{-t_v/t^*\right\}}{1 + \exp\left\{-t_v/t^*\right\}}. \tag{89}$$

can be named by the effective coefficient of the elastic component of the viscosity of polymeric melt or, in general case, by the single-component macromolecular liquid.

In accordance with the Eqs. (84)–(86) the following conditions should be performed:

$$g \to 0,\; t_v \to \infty \;\text{and}\; g \to \infty,\; t_v \to 0, \tag{90}$$

at which from the Eq. (87) two asymptotic limits for the effective viscosity respectively follow:

$$\eta = \eta_f + \eta_e^0, \tag{91}$$

and

$$\eta = \eta_f. \tag{92}$$

So, accordingly to Eq. (91) a so-called *"maximal Newton viscosity"* [15] at $g \to 0$ is not as such, since is represented by a sum of the frictional *(Newton)* and of the elastic *(Maxwell)* components. On the contrary, at $g \to \infty$ the shear deformation of the conformational volume of polymeric chain hasn't time is visualized and the endowment of the elastic component of the viscosity becomes as zero; so, the effective viscosity is wholly determined by the frictional component.

Due to the volumetric character of the shear forces the Eq. (40) for the polymeric solution as a function of its composition is retained in the previous form

$$\eta = \sum_i \eta_i \phi_i \, . \tag{93}$$

But at this, the viscosity coefficient of polymer in Eq. (93) should be presented with taking into account the frictional and elastic components that is in a form of Eq. (87). Besides, it is necessary to take into account that the properties of polymeric chains, in particular, the module μ and the characteristic time t^* of shear, essentially depend on fact, if the polymeric solution is diluted or concentrated accordingly to the conditions $\rho \le \rho^*$ or $\rho > \rho^*$, in which ρ is the density, and ρ^* is the critical density of the solution upon polymer, which corresponds to the beginning of the polymeric chains conformational volumes overlapping.

These and others nuances of the Eq. (93) application to the analysis of the effective viscosity of polymeric solutions will be in detail discussed in two following articles with the using of more statistically significant experimental data concerning to the viscosity of the polystyrene solutions and melts.

5.7 CONCLUSIONS

An analysis of the molecular-kinetic theories of *Frenkel* and *Eyring* leads to the expressions for the coefficients of viscosity of low-molecular pure liquids, containing of the ratio $RT\tau/V$. From the general phenomenological determinations of the substances flow under the action of the gradient of chemical potential and analysis of the appearing into the flow of the shear forces and corresponding tensions it was proved, that the viscosity coefficient of the pure liquid is ordered to the ratio $\eta = 3RT\tau/V$, and in a case of the component of solution it is ordered to the ratio $\eta_i = 3RT\tau_i / \overline{V}_i$.

A comparison of the experimental values of the characteristic times t of the viscous flow and the calculated ones shows, that the preexponential multiplier τ_0 is determined by not only the frequency of the oscillating movement of the particles into the quasi-lattice of the liquid but also by the entropy factor. This leads to the conclusion that the activation entropy at the viscous flow of the liquid can be found via the same expression, as in a case of the entropy at the phase transition by the first kind. Obtained expression for the activation entropy $\Delta S^* = \Delta H^* / T^*$ permits to explain the low values of $\tau_0 \ll 2h/kT$ for the associated liquids and the observed slope opposition for the dependence between τ_0 and the activation energy of the viscous flow.

It were proposed the expressions accordingly to which it can be calculated the coefficients of the self-diffusion and diffusion based on corresponding coefficients of the viscosity of low-molecular pure liquids and melts.

An analysis of the *Maxwell's* equation and the deformation rates of the conformational volumes of polymeric chains and their rotation permitted to separate the frictional and elastic coefficients of the viscosity of high-molecular single-component liquid. It was shown, that exactly the elastic coefficient of the viscosity is gradiently depended value. At this, a so-called maximal *Newton* viscosity at $g \to 0$ indeed is not such and it is represented by a sum of the frictional and elastic components. On the contrary, at $g \to \infty$ the effective viscosity is wholly determined only by the frictional component.

KEYWORDS

- **characteristic time of the viscous flow**
- **chemical potential**
- **frictional and elastic components of the viscosity of high molecular liquid**
- **viscosity of pure liquids**

REFERENCES

1. Frenkel J., *"Z. Phys."*, **35**, 652 (1926).
2. Frenkel Ya. I., Kinetic Theory of Liquid, *"Science Edition"*, Leningrad (1975) *(in Russian)*.
3. Eyring H., *"J. Chem. Phys."*, **4**, 283 (1936).
4. Ewel, R. H., Eyiring, H. *"J. Chem. Rhys."*, **5**, 726 (1937).
5. Glesston S., Leidler K., Eiring H. Theory of Absolute Reactions Rates, *"Scientific Literature Edition"*, Moscow (1948) *(in Russian)*.
6. Kendall J., Monroe L. P., *"J. Amer. Chem. Soc."*, **39**, 1789 (1917).
7. Lonsdale H. K., Marten U., Rilly R. L., *"J. Appl. Polymer. Sci."*, **9**, 1341 (1965).
8. Medvedevskikh Yu. G., Turovsky A. A., Zaikov G. E., *"Chem. Phys."*, **17** (8), 141 (1998) *(in Russian)*.
9. Medvedevskikh Yu. G., Kytsya A. R., Bazylyak L. I., Turovsky A. A., Zaikov G. E., Stationary and Non-Stationary Kinetics of Photoinitiated Polymerization, *"VSP Leiden Edition"*, Boston, 307 p. (2004).
10. Medvedevskikh Yu. G., Kytsya A. R., Bazylyak L. I., Zaikov G. E., In: *"Conformation of Macromolecules"*, Editors: Yu. G. Medvedevskikh et al. / Nova Science Publishers, Inc., 145–157 (2007).
11. Medvedevskikh Yu. G., *"Cond. Matt. Phys."*, **4** (2)/(26), 209, 219 (2001).
12. Medvedevskikh Yu. G., *"J. Appl. Pol. Sci."*, **109** (4), 2472 (2008).
13. Kuhn Y., Kuhn W., *"J. Polymer. Sci."*, **5**, 519 (1952); **9**, 1 (1950).
14. de Gennes P. G. Skaling Concepts in Polymer Physics, *"Cornell Univ. Press."* (1979).
15. Malkin A. Ya., Isayev A. I., Rheology: Conceptions, Methods, Applications, *"Tec. Publishing"*, Toronto, Canada (2005) *(in Russian)*.

REFERENCES

1.
2.
3.
4.
5.
6.
7.
8.
9.
10.
11.
12.
13.

CHAPTER 6

VISCOELASTIC PROPERTIES OF THE POLYSTYRENE IN CONCENT-RATED SOLUTIONS AND MELTS

YU. G. MEDVEDEVSKIKH, O. YU. KHAVUNKO, L. I. BAZYLYAK, and G. E. ZAIKOV

CONTENTS

SUMMARY

A gradient dependence of the effective viscosity η for the concentrated solutions of the polystyrene in toluene at three concentrations $\rho = 0.4 \times 10^5$; 0.5×10^5; 0.7×10^5 g/m^3 correspondingly for the fourth fractions of the polystyrene with the average molar weights $M = 5.1 \times 10^4$; 4.1×10^4; 3.3×10^4; 2.2×10^4 g/mole respectively has been experimentally investigated. For every pair of the values ρ and M a gradient dependence of the viscosity was studied at four temperatures: 25, 30, 35 and 40°C. An effective viscosity of the melts of polystyrene was studied for the same fractions, but at the temperatures 190, 200 and 210 °C. The investigations have been carried out with the use of the rotary viscosimeter "Rheotest 2.1" under the different angular velocities ω of the working cylinder rotation. An analysis of the dependencies $\eta(\omega)$ permitted to mark the frictional η_f and elastic η_e components of the viscosity ant to study their dependence on temperature T, concentration ρ and on the length of a chain N. It was determined, that the relative movement of the intertwined between themselves polymeric chains into m-ball, which includes into itself the all possible effects of the gearings, makes the main endowment into the frictional component of the viscosity. The elastic component of the viscosity η_e is determined by the elastic properties of the conformational volume of the m-ball of polymeric chains under its shear strain. The numerical values of the characteristic time and the activation energy of the segmental movement were obtained on the basis of the experimental data. In a case of a melt the value of E and $\Delta S^*/R$ are approximately in two times more than the same values for the diluted and concentrated solutions of the polystyrene in toluene; this means that the dynamic properties of the polymeric chains in melt are considerably near to their values in polymeric matrix than in solutions. Carried out analysis and generalization of the obtained experimental data show that as same as for low-molecular liquids the studying of the viscosity of polymeric solutions permits sufficient adequate to estimate the characteristic time of the segmental movement accordingly to which the coefficients of polymeric chains diffusion can be calculated in solutions and melt, in other words, to determine their dynamic characteristics.

6.1 INTRODUCTION

The viscosity h of polymeric solutions is an object of the numerous experimental and theoretical investigations generalized in Refs. [1-4]. This is explained both by the practical importance of the presented property of polymeric solutions in a number of the technological processes and by the variety of the factors having an influence on the h value, also by a wide diapason (from 10^{-3} to 10^{2} $Pa.s$) of the viscosity change under transition from the diluted solutions and melts to the concentrated ones. The all above said gives a great informational groundwork for the testing of different theoretical imaginations about the equilibrium and dynamic properties of the polymeric chains.

It can be marked three main peculiarities for the characteristic of the concentrated polymeric solutions viscosity, *namely*:

1. Measurable effective viscosity h for the concentrated solutions is considerable stronger than the h for the diluted solutions and depends on the velocity gradient g of the hydrodynamic flow or on the shear rate.

It can be distinguished [4] the initial η_0 and the final η_∞ viscosities ($\eta_0 > \eta_\infty$), to which the extreme conditions $g \rightarrow 0$ and $g \rightarrow \infty$ correspond respectively.

Due to dependence of η on g and also due to the absence of its theoretical description, the main attention of the researches [4] is paid into, so-called, the most newton (initial) viscosity η_0, which is formally determined as the limited value at $g \rightarrow 0$. Exactly this value η_0 is estimated as a function of molar mass, temperature, and concentration (in solutions).

The necessity of the experimentally found values of effective viscosity extrapolation to "zero" shear stress doesn't permit to obtain the reliable value of η_0. This leads to the essential and far as always easy explained contradictions of the experimental results under the critical comparison of data by different authors.

2. Strong power dependence of h on the length N of a polymeric chain and on the concentration r (g/m^3) of a polymer in solution exists: $\eta \sim \rho^\alpha N^\beta$ with the indexes $\alpha = 5$, 7, $b = 3.3$, 3.5, as it was shown by authors [4].

3. It was experimentally determined by authors [1, 5] that the viscosity h and the characteristic relaxation time t^* of the polymeric chains into concentrated solutions and melts are characterized by the same scaling dependence on the length of a chain:

$$\eta \sim t^* \sim N^\beta \tag{1}$$

with the index $b = 3.4$.

Among the numerous theoretical approaches to the analysis of the polymeric solutions viscosity anomaly, that is, the dependence of h on g, it can be marked the three main approaches. The first one connects the anomaly of the viscosity with the influence of the shear strain on the potential energy of the molecular kinetic units transition from the one equilibrium state into another one and gives the analysis of this transition from the point of view of the absolute reactions rates theory [6]. However, such approach hasn't take into account the specificity of the polymeric chains; that is why, it wasn't win recognized in the viscosity theory of the polymeric solutions. In accordance with the second approach the polymeric solutions viscosity anomaly is explained by the effect of the hydrodynamic interaction between the links of the polymeric chain; such links represent by themselves the "beads" into the "necklace" model. Accordingly to this effect the hydrodynamic flow around the presented "bead" essentially depends on the position of the other "beads" into the polymeric ball. An anomaly of the viscosity was conditioned by the anisotropy of the hydrodynamic interaction, which creates the orientational effect [7, 8]. High values of the viscosity for the concentrated solutions and its strong gradient dependence cannot be explained only by the effect of the hydrodynamic interaction.

That is why the approaches integrated into the conception of the structural theory of the viscosity were generally recognized. In accordance with this theory the viscosity of the concentrated polymeric solutions is determined by the quasi-net of the linkages of twisted between themselves polymeric chains and, therefore, depends on the modulus of elasticity E of the quasi-net and on the characteristic relaxation time t^* [1–2]:

$$\eta = E \cdot t^* \tag{2}$$

It is supposed, that the E is directly proportional to the density of the linkages assemblies and is inversely proportional to the interval between them along the same chain. An anomaly of the viscosity is explained by the linkages assemblies' density decreasing at their destruction under the action of shear strain [9], or by the change of the relaxation spectrum [10], or by the distortion of the polymer chain links distribution function relatively to its center of gravity [11]. A gradient dependence of the viscosity is described by the expression [11]:

$$(\eta - \eta_\infty)/(\eta_0 - \eta_\infty) = f\left(gt^*\right) \tag{3}$$

It was greatly recognized the universal scaling ratio [1, 5]:

$$\eta = \eta_0 \cdot f\left(gt^*\right) \tag{4}$$

in which the dimensionless function $f\left(gt^*\right) = f(x)$ has the asymptotes $f(0)$ $= 1, f(x)_{x>1} = x^{-g}, g = 0.8$.

Hence, both Eqs. (3) and (4) declare the gradient dependence of h by the function of the one nondimensional parameter gt^*. However, under the theoretical estimation of h and t^* as a function of N there are contradictions between the experimentally determined ratio (1) and $b = 3.4$. Thus, the analysis of the entrainment of the surrounding chains under the movement of some separated chain by [12] leads to the dependencies $\eta \sim N^{3.5}$ but $t^* \sim N^{4.5}$. At the analysis [13] of the self-coordinated movement of a chain enclosing into the tube formed by the neighboring chains it was obtained the $\eta \sim N^3$, $t^* \sim N^4$. The approach in Ref. [14], which is based on the conception of the reputational mechanism of the polymeric chain movement gives the following dependence $\eta \sim t^* N^3$. So, the index $b = 3.4$ in the ratio (Eq. 1) from the point of view of authors [2] remains by one among the main unsolved tasks of the polymers' physics.

Summarizing the above presented short review, let us note, that the conception about the viscosity-elastic properties of the polymeric solutions accordingly to the *Maxwell's* equation should be signified the presence of two components of the effective viscosity, *namely*: the *frictional* one, caused by the friction forces only, and the *elastic* one, caused by the shear strain of the conformational volume of macromolecules. But in any among listed above theoretical approaches the shear strain of the conformational volumes of macromolecules was not taken into account. The sustained opinion by authors [3-4] that the shear strain is visualized only in the strong hydrodynamic flows whereas it can be neglected at little g, facilitates to this fact. But in this case the inverse effect should be observed, *namely* an increase of h at the g enlargement.

These contradictions can be overpassed, if to take into account [15, 16], that, although at the velocity gradient of hydrodynamic flow increasing the external action leading to the shear strain of the conformational volume of polymeric chain is increased, but at the same time, the characteristic time of the external action on the rotating polymeric ball is decreased; in accordance with the kinetic reasons this leads to the decreasing but not to the increasing of the shear strain degree. Such analysis done by authors [15-17] permitted to mark the *frictional* and the *elastic* components of the viscosity and to show that *exactly the elastic component* of the viscosity *is the gradiently dependent value*. The elastic properties of the conformational volume of polymeric chains, in particular shear modulus, were described early by authors [18-19] based on the self-avoiding walks statistics (*SAWS*).

Here presented the experimental data concerning to the viscosity of the concentrated solutions of styrene in toluene and also of the melt and it is given their interpretation on the basis of works [15-19].

6.2 EXPERIMENTAL DATA AND STARTING POSITIONS

In order to obtain statistically significant experimental data we have studied the gradient dependence of the viscosity for the concentrated solution of polystyrene in toluene at concentrations 0.4×10^5; 0.5×10^5 and 0.7×10^5 g/m^3 for the four fractions of polystyrene characterizing by the apparent molar weights $M = 5.1 \times 10^4$; $M = 4.1 \times 10^4$; $M = 3.3 \times 10^4$ and $M = 2.2 \times 10^4$ $g/mole$. For each pair of values r and M the gradient dependence of the viscosity has been studied at fourth temperatures 25°C, 30°C, 35°C and 40°C.

The viscosity for the polystyrene melt were investigated using the same fractions at 210°C. Temperature dependence of the polystyrene melt was investigated for the fraction with average molecular weight $M = 2.2 \times 10^4$ g/$mole$ under three temperatures, namely 190, 200 and 210°C.

The experiments have been carried out with the use of the rotary viscometer $RHEOTEST$ 2.1 equipped by the working cylinder having two rotary surfaces by diameters $d_1 = 3.4 \times 10^{-2}$ and $d_2 = 3.9 \times 10^{-2}$ m in a case of the concentrated solutions of polystyrene investigation and using the device by "cone–plate" type equipped with the working cone by 0.3° angle and radius $r = 1.8 \times 10^{-2}$ m in a case of the polystyrene melt investigation.

6.3 CONCENTRATED SOLUTIONS

6.3.1 INITIAL STATEMENTS

Typical dependences of viscosity η of solution on the angular velocity w *(turns/s)* of the working cylinder rotation are represented on Figs. 1–3. Generally it was obtained the 48 curves of $h(w)$.

For the analysis of the experimental curves of $h(w)$ it was used the expression [15, 20]:

$$\eta = \eta_f + \eta_e \left(1 - \exp\{-b/\omega\}\right) / \left(1 + \exp\{-b/\omega\}\right) \tag{5}$$

in which η is the measured viscosity of the solution at given value ω of the working cylinder velocity rate; η_f and η_e are frictional and elastic components of η;

$$b / \omega = t_v^* / t_m^* \tag{6}$$

where t_m^* is the characteristic time of the shear strain of the conformational volume for m-ball of intertwined polymeric chains; t_v^* is the characteristic time of the external action of gradient rate of the hydrodynamic flow on the m-ball.

The notion about the m-ball of the intertwined polymeric chains will be considered later.

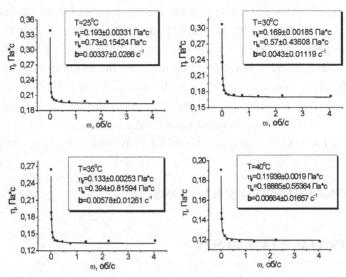

FIGURE 1 Experimental (points) and calculated in accordance with the Eq. (5) (curves) dependencies of the effective viscosity on the rotation velocity of the working cylinder: ρ = 4.0×10⁵ g/m³, M = 4.1×10⁴ g/m*ole*, T = 25 ÷ 40°C.

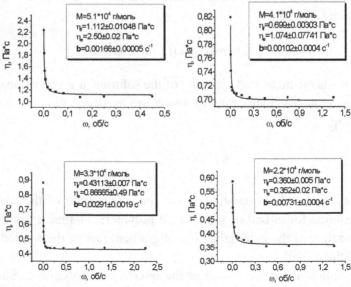

FIGURE 2 Experimental (points) and calculated in accordance with the Eq. (5) (curves) dependencies of the effective viscosity on the rotation velocity of the working cylinder: ρ = 5.0×10⁵ g/m³, M = 5.1 ÷ 2.2×10⁴ g/m*ole*, T = 25°C.

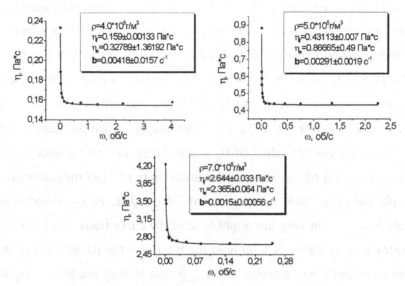

FIGURE 3 Experimental (points) and calculated in accordance with the Eq. (5) (curves) dependencies of the effective viscosity on the rotation velocity of the working cylinder: ρ = 4.0×10⁵ , 7.0×10⁵ g/m³, $M = 3.3×10^4$ g/mole, $T = 25°C$.

The shear strain of the conformational volume of m-ball and its rotation is realized in accordance with the reputational mechanism presented in Ref. [2], that is, via the segmental movement of the polymeric chain, that is why t_m^* is also the characteristic time of the own, that is, without the action g, rotation of m-ball [17].

The Eq. (5) leads to the two asymptotes:

$$\eta = \eta_f + \eta_e \text{ at } b/\omega > 1$$

$$\eta = \eta_f \text{ at } b/\omega < 1$$

So, it is observed a general regularity of the effective viscosity dependence on the rotation velocity ω of the working cylinder for diluted, concentrated solutions and melts. Under condition, that $b/\omega > 1$, that is at $\omega \to 0$, the effective viscosity is equal to a sum of the frictional and elastic components of the viscosity, and under condition $\omega \to \infty$ the measurable viscosity is determined only by a frictional component of the viscosity.

In accordance with Eq. (5) the effective viscosity $h(w)$ is a function on three parameters, *namely* η_f, η_e and b. They can be found on a basis of the experimental values of $h(w)$ via the optimization method in program *ORIGIN* 5.0. As an analysis showed, the numerical values of η_f are easy determined upon a plateau on the curves $h(w)$ accordingly to the condition $b/\omega < 1$ (*see* Figs. 1–3). However, the optimization method gave not always the correct values of η_e and b. There are two reasons for this. Firstly, in a field of the $\omega \to 0$, the uncertainty of $h(w)$ measurement is sharply increased since the moment of force registered by a device is a small. Secondly, in very important field of the curve transition $h(w)$ from the strong dependence of h on w to the weak one the parameters η_e and b are interflowed into a composition $\eta_e b$, that is, they are by one parameter. Really, at the condition $b/\omega < 1$ decomposing the exponents into Eq. (5) and limiting by two terms of the row $exp\left\{-\dfrac{b}{\omega}\right\} \approx 1 - \dfrac{b}{\omega}$, we will obtained $\eta = \eta_f + \eta_e b/2$. Due to the above-mentioned reasons the optimization method gives the values of η_e and b depending between themselves but doesn't giving the global minimum of the errors functional. That is why at the estimation of η_e and b parameters it was necessary sometimes to supplement the optimization method with the "manual" method of the global minimum search varying mainly by the numerical estimation of η_e.

As we can see from the Figs. 1–3, calculated curves $\eta(\omega)$ accordingly to the Eq. (5) and found in such a way parameters η_f, η_e and b, are described the experimental values very well.

The results of η_f, η_e and b numerical estimations for the all-48 experimental curves $\eta(\omega)$ are represented in Table 1. The mean−square standard deviations of the η_f, η_e and b calculations indicated on the Figs. 1–3.

A review of these data shows that the all three parameters are the functions on the concentration of polymer into solution, on the length of a chain and on the temperature. But at this, the η_e and η_f are increased at the r and M increasing and are decreased at the T increasing whereas the b parameter is changed into the opposite way. The analysis of these dependencies will be represented in Chapter 3 (Section 3.2 and 3.3). Here let us present the all needed for this analysis determinations, notifications and information concerning to the concentrated polymeric solutions.

Investigated solutions of the polystyrene in toluene were concentrated; since the following condition was performing for them:

$$\rho \geq \rho^* \tag{7}$$

where ρ^* is a critical density of the solution per polymer corresponding to the starting of the polymeric chains conformational volumes overlapping having into diluted solution $(\rho \leq \rho^*)$ the conformation of *Flory* ball by the radius

$$R_f = aN^{3/5}, \tag{8}$$

where a is a length of the chain's link. It's followed from the determination of ρ^*

$$\rho^* = M / N_A R_f^3 = M_0 N / N_A R_f^3 \tag{9}$$

where M_0 is the molar weigh of the link of a chain. Taking into account the Eqs. (8) and (9) we have:

$$\rho^* = \rho_0 N^{-4/5}, \tag{10}$$

where

$$\rho_0 = M_0 / a^3 N_A \tag{11}$$

can be called as the density into volume of the monomeric link.

TABLE 1   Optimization parameters η, η_e and b in Eq. (5).

$\rho \times 10^{-5}$, g/m³		4.0				5.0				7.0		
T , °C	M×10⁴ g/mole	5.1	4.1	3.3	2.2	5.1	4.1	3.3	2.2	5.1	4.1	3.3	2.2

	$\eta'\times$, Pa$\times$s	0.35	0.19	0.16	0.06	1.11	0.69	0.43	0.36	6.50	2.66	2.64	0.86
25	η_e Pa$\times$s	1.40	0.73	0.33	0.09	2.50	1.10	0.87	0.35	7.60	3.75	2.37	1.50
	$b\times10^3$, s^{-1}	1.15	3.37	4.20	32.3	1.66	1.02	2.91	7.31	0.36	0.76	1.50	2.44
	$\eta'\times$, Pa$\times$s	0.31	0.17	0.14	0.05	1.00	0.62	0.36	0.24	4.95	2.11	2.03	0.68
30	η_e Pa$\times$s	0.95	0.57	0.25	0.06	1.30	0.76	0.52	0.32	4.05	2.21	1.86	1.00
	$b\times10^3$, s^{-1}	1.38	4.30	5.90	35.0	2.23	1.80	3.14	8.69	0.72	0.83	1.70	2.65
	$\eta'\times$, Pa$\times$s	0.19	0.13	0.11	0.04	0.68	0.50	0.26	0.19	4.07	1.85	1.45	0.43
35	η_e Pa$\times$s	0.60	0.39	0.21	0.05	0.90	0.35	0.23	0.22	3.50	1.80	1.59	0.79
	$b\times10^3$, s^{-1}	3.67	5.80	6.37	49.0	2.41	3.56	4.60	9.10	0.88	0.96	1.93	3.20
	$\eta'\times$, Pa$\times$s	0.17	0.12	0.10	0.04	0.56	0.42	0.22	0.17	2.91	1.46	0.98	0.27
40	η_e Pa$\times$s	0.40	0.19	0.13	0.03	0.65	0.29	0.15	0.12	2.01	1.39	1.19	0.57
	$b\times10^3$, s^{-1}	5.35	6.60	6.90	73.9	2.67	5.60	5.60	16.8	1.33	1.41	2.27	4.24

In accordance with the *SARWS* [19] the conformational radius R_m of the polymeric chain into concentrated solutions is greater than into diluted ones and is increased at the polymer concentration ρ increasing. Moreover, not one, but m macromolecules with the same conformational radius are present into the conformational volume R_m^3. This leads to the notion of twisted polymeric chains m-ball for which the conformational volume R_m^3 is general and equally accessible. Since the m-ball is not localized with the concrete polymeric chain, it is the virtual, that is, by the mathematical notion.

It is followed from the *SARWS* [19]:

$$R_m = R_f \cdot m^{1/5} \tag{12}$$

$$m^{1/5} = \left(\rho/\rho^*\right)^{1/2} \text{ at } \rho \ge \rho^*, \tag{13}$$

thus, it can be written

$$R_m = aN(\rho/\rho_0)^{1/2}$$

(14)

The shear modulus μ for the m-ball was determined by the expression [19]:

$$\mu = 1.36\frac{RT}{N_A a^3}\left(\frac{\rho}{\rho_0}\right)^2$$

(15)

and, as it can be seen, doesn't depend on the length of a chain into the concentrated solutions.

Characteristic time t_m^* of the rotary movement of the m-ball and, respectively its shear, in accordance with the prior work [17] is equal to

$$t_m^* = \frac{4}{7}N^{3.4}\left(\frac{\rho}{\rho_0}\right)^{2.5}L_m\tau_m$$

(16)

Let us compare the t_m^* with the characteristic time t_f^* of the rotary movement of *Flory* ball into diluted solution [17]:

$$t_f^* = \frac{4}{7}N^{1.4}L_f\tau_f.$$

(17)

In these expressions τ_m and τ_f are characteristic times of the segmental movement of the polymeric chains and L_m and L_f are their form factors into concentrated and diluted solutions respectively. Let us note also, that the Eqs. (16) and (17) are self-coordinated since at $\rho = \rho^*$ the Eq. (16) transforms into the Eq. (17). The form factors L_m and L_f are determined by a fact how much strong the conformational volume of the polymeric chain is strained into the ellipsoid of rotation, flattened or elongated one as it was shown by author [21].

6.3.2 FRICTIONAL COMPONENT OF THE EFFECTIVE VISCOSITY

In accordance with the data of Table 1 the frictional component of the viscosity η_f strongly depends on a length of the polymeric chains, on their concentration and on the temperature. The all spectrum of η_f dependence on N, ρ and T we will be considered as the superposition of the fourth movement forms giving the endowment into the frictional component of the solution viscosity. For the solvent such movement form is the *Brownian* movement of the molecules, i.e., their translation freedom degree: the solvent viscosity coefficient η_s will be corresponding to this translation freedom degree. The analog of the *Brownian* movement of the solvent molecules is the segmental movement of the polymeric chain, which is responsible for its translation and rotation movements and also for the shear strain. The viscosity coefficient η_{sm} will be corresponding to this segmental movement of the polymeric chain.

Under the action of a velocity gradient g of the hydrodynamic flow the polymeric m-ball performs the rotary movement also giving the endowment into the frictional component of the viscosity. In accordance with the superposition principle the segmental movement and the external rotary movement of the polymeric chains will be considered as the independent ones. In this case the external rotary movement of the polymeric chains without taking into account the segmental one is similar to the rotation of m-ball with the frozen equilibrium conformation of the all m polymeric chains represented into m-ball. This corresponds to the inflexible *Kuhn's* wire model [22]. The viscosity coefficient η_{pm} will be corresponding to the external rotating movement of the m-ball under the action of g. The all listed movement forms are enough in order to describe the diluted solutions. However, in a case of the concentrated solutions it is necessary to embed one more movement form, *namely*, the transference of the twisted between themselves polymeric chain one respectively another in m-ball. Exactly such relative movement of the polymeric chains contents into itself the all-possible linkages effects. Accordingly to the superposition principle the polymeric chains movement does not depend on the above-listed movement forms if it doesn't change the equilibrium conformation of the polymeric chains in m-ball. The endowment of such movement form into η_f let us note via η_{pz}.

Not all the listed movement forms give the essential endowment into the η_f, however for the generality let us start from the taking into account of the all forms. In such a case the frictional component of a viscosity should be described by the expression:

$$\eta_f = \eta_s(1-\phi)+\left(\eta_{sm}+\eta_{pm}+\eta_{pz}\right)\phi,\qquad(18)$$

or

$$\eta_f - \eta_s =\left(\eta_{sm}+\eta_{pm}+\eta_{pz}-\eta_s\right)\phi,\qquad(19)$$

where φ is the volumetric part of the polymer into solution. It is equal to the volumetric part of the monomeric links into m-ball; that is why it can be determined by the ratio:

$$\phi=\bar{V}N/N_A R_m^3,\qquad(20)$$

in which $\bar{V}$ is the partial-molar volume of the monomeric link into solution.

Combining the Eqs. (9)–(14) and Eq. (20) we will obtain:

$$\phi=\bar{V}\rho/M_0.\qquad(21)$$

The ratio of $M_0/\bar{V}$ should be near to the density ρ_m of the liquid monomer. Assuming of this approximation, $M_0/\bar{V}\approx\rho_m$ we have:

$$M_0/\bar{V}\approx\rho_m.\qquad(22)$$

At the rotation of m-ball under the action of g the angular rotation rate for any polymeric chain is the same but their links depending on the remoteness from the rotation center will have different linear movement rates. Consequently, in m-ball there are local velocity gradients of the hydrodynamic flow. Let g_m represents the averaged upon m-ball local velocity gradient of the hydrodynamic flow additional to g. Then, the tangential or strain shear σ formed by these gradients g_m and g at the rotation movement of m-ball in the medium of a solvent will be equal to:

$$\sigma = \eta_s \left(g + g_m \right).$$

(23)

However, the measurable strain shear correlates with the well-known external gradient g that gives another effective viscosity coefficient:

$$\sigma = \eta_{pm} g$$

(24)

Comparing the Eqs. (23) and (24) we will obtain

$$\eta_{pm} - \eta_s = \eta_s g_m / g.$$

(25)

Noting

$$\eta_{pm}^0 = \eta_s \cdot g_m / g$$

(26)

instead of the Eq. (19) we will write

$$\eta_f - \eta_s = \left(\eta_{sm} + \eta_{pm}^0 + \eta_{pz} \right) \phi$$

(27)

The endowment of the relative movement of twisted polymeric chains in m-ball into the frictional component of the viscosity should be in general case depending on a number of the contacts between monomeric links independently to which polymeric chain these links belong. That is why we assume:

$$\eta_{pz} \sim \phi^2.$$

(28)

The efficiency of these contacts or linkages let us estimate comparing the characteristic times of the rotation (shear) of m-ball into concentrated solution t_m^* and polymeric ball into diluted solution t_f^* determined by the Eqs. (16) and (17).

Let's note that in accordance with the determination done by author [17] t_m^* is the characteristic time not only for m-ball rotation, but also for each polymeric chain in it. Consequently, t_m^* is the characteristic time of the rotation of polymeric chain twisted with others chains whereas t_f^* is the characteristic time of free polymeric chain rotation. The above-said permits to assume the ratio t_m^* / t_f^* as a measure of the polymeric chains contacts or linkages efficiency and to write the following in accordance with the Eqs. (16) and (17):

$$\eta_{pz} \sim t_m^* / t_f^* = N^2 (\rho/\rho_0)^{2.5} (L_m/L_f) \tag{29}$$

Taking into account the Eq. (22) and combining the Eqs. (28) and (29) into one expression we will obtain:

$$\eta_{pz} = \eta_{pz}^0 N^2 \left(\frac{\rho}{\rho_0}\right)^{2.5} \left(\frac{\rho}{\rho_m}\right)^2 \tag{30}$$

Here the coefficient of proportionality η_{pz}^0 includes the ratio $L_m \tau_m / L_f \tau_f$, which should considerably weaker depends on ρ and N that the value η_{pz}.

Substituting the Eqs. (30) into (27) with taking into account the Eq. (22) we have:

$$\eta_f - \eta_s = \left[\eta_{sm} + \eta_{pm}^0 + \eta_{pz}^0 N^2 \left(\frac{\rho}{\rho_0}\right)^{2.5} \left(\frac{\rho}{\rho_m}\right)^2 \right] \frac{\rho}{\rho_m} \tag{31}$$

Let us estimate the endowment of the separate terms in Eq. (31) into η_f. In accordance with Table 1 under conditions of our experiments the frictional component of the viscosity is changed from the minimal value » 4×10^{-2} $Pa \cdot s$ to the maximal one » 6.5 $Pa \cdot s$. Accordingly to the reference data the viscosity coefficient η_s of the toluene has the order 5×10^{-4} $Pa \cdot s$. The value of the viscosity coefficient η_{sm} representing the segmental movement of the polymeric chains estimated by us upon η_f of the diluted solution of polystyrene in toluene consists of the value by 5×10^{-3} $Pa \cdot s$ order. Thus, it can be assumed $\eta_{sm}, \eta_s < \eta_f$ and it can be neglected the respective terms in Eq. (31). With taking into account of this fact, the Eq. (31) can be rewritten in a form convenient for the graphical test:

$$\eta_f \frac{\rho_m}{\rho} = \eta_{pm}^0 + \eta_{pz}^0 N^2 \left(\frac{\rho}{\rho_0}\right)^{2.5} \left(\frac{\rho}{\rho_m}\right)^2. \tag{32}$$

On Fig. 4, it is presented the interpretation of the experimental values of η_f into coordinates of the Eq. (32).

At that, it were assumed the following values: $M = 104.15$ g/mole, $a = 1.86 \times 10^{-10}$ m under determination of ρ_0 accordingly to Eq. (11) and $\rho_m = 0,906 \cdot 10^6$ g/m^3 for liquid styrene. As we can, the linear dependence is observed corresponding to Eq. (32) at each temperature; based on the tangent of these straight lines inclination (see the regression equations on Fig. 4) it were found the numerical values of η_{pz}^0, the temperature dependence of which is shown on Fig. 5 into the *Arrhenius'* coordinates.

It is follows from these data, that the activation energy E_{pz} regarding to the movement of twisted polymeric chains in toluene is equal to 39.9 kJ/mole.

It can be seen from the Fig. 4 and from the represented regression equations on them, that the values η_{pm}^0 are so little (probably, $\eta_{pm}^0 \ll 0,1$ *Pa·s*) that they are located within the limits of their estimation error. This, in particular, didn't permit us to found the numerical values of the ratio g_m / g.

So, the analysis of experimental data, which has been done by us, showed that the main endowment into the frictional component of the effective viscosity of the concentrated solutions "polystyrene in toluene" has the separate movement of the twisted between themselves into m-ball polymeric chains. Exactly this determines a strong dependence of the η_f on concentration of polymer into solution $\left(\eta_f \sim \rho^{5,5}\right)$ and on the length of a chain $\left(\eta_f \sim N^2\right)$.

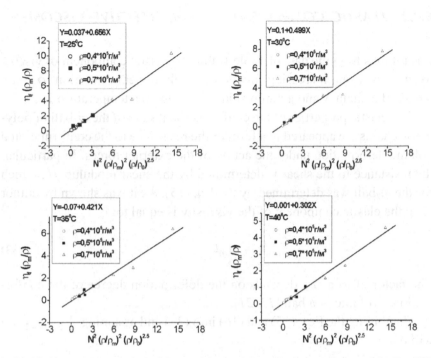

FIGURE 4   An interpretation of the experimental data of η_f in coordinates of the Eq. (32).

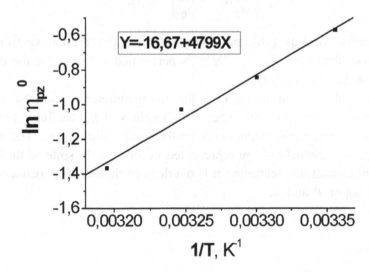

FIGURE 5   Temperature dependence of the viscosity coefficient η_{pz}^0 in coordinates of the *Arrhenius* equation.

6.3.3 ELASTIC COMPONENT OF THE EFFECTIVE VISCOSITY

It is follows from the data of Table 1, that the elastic component of viscosity η_e is a strong increasing function on polymer concentration ρ, on a length of a chain N and a diminishing function on a temperature T.

The elastic properties of the conformational state of the m-ball of polymeric chains are appeared in a form of the resistance to the conformational volume deformation under the action of the external forces. In particular, the resistance to the shear is determined by the shear modulus μ, which for the m-ball was determined by the Eq. (15). As it was shown by author [17], the elastic component of the viscosity is equal to:

$$\eta_e = \mu t_m^* L_m. \tag{33}$$

The factor of form L_m depends on the deformation degree of the conformational volume of a ball [17, 21].

Combining the Eqs. (15) and (16) into (33) and assuming $\frac{4}{7} \cdot 1,36 \approx 1$ we will obtain

$$\eta_e = \frac{RT}{M_0} N^{3.4} \rho \left(\frac{\rho}{\rho_0} \right)^{3.5} L_m \tau_m. \tag{34}$$

Comparing the Eqs. (16) and (34) we can see that the known from the reference data ratio $\eta_e \sim t_m^* \sim N^{3.4}$ is performed but only for the elastic component of a viscosity.

It is follows from the Eq. (34), that the parameters L_m and τ_m are inseparable; so, based on the experimental values of η_e (see Table 1) it can be found the numerical values only for the composition $L_m \cdot \tau_m$. The results of $(L_m \tau_m) \eta$ calculations are represented in Table 2. In spite of these numerical estimations scattering it is overlooked their clear dependence on T, but not on ρ and N.

TABLE 2   Calculated values $L\tau$, τ/L, τ and L based on the experimental magnitudes η_e and b.

T, °C	M×10⁴, g/mole	ρ×10⁻⁵, g/m³ 4.0				5.0				7.0				τ10¹⁰, s L
		5.1	4.1	3.3	2.2	5.1	4.1	3.3	2.2	5.1	4.1	3.3	2.2	
25	$(L\tau)_{\eta e}\times10^{10}$, s	2.63	3.14	2.72	2.99	1.71	1.72	2.61	4.25	1.15	1.29	1.57	4.00	
	$(\tau/L)_b\times10^{10}$, s	3.25	1.81	2.54	0.89	1.17	3.43	1.91	2.06	1.98	1.86	1.38	2.29	
	$\tau\times10^{10}$, s	2.92	2.38	2.63	1.63	1.41	2.43	2.23	2.96	1.51	1.61	1.47	3.03	2.19
	L	0.90	1.32	1.03	1.83	1.21	0.71	1.17	1.44	0.76	0.86	1.07	1.32	1.13
30	$(L\tau)_{\eta e}\times10^{10}$, s	1.75	2.41	2.03	1.96	0.88	1.17	1.54	3.83	0.60	0.75	1.21	2.63	
	$(\tau/L)_b\times10^{10}$, s	2.10	1.56	1.81	0.82	0.87	1.94	1.39	1.73	1.00	1.62	1.22	2.11	
	$\tau\times10^{10}$, s	2.17	1.94	1.92	1.27	0.88	1.51	1.46	2.57	0.78	0.98	1.21	2.56	1.59
	L	0.81	1.24	1.00	1.55	1.00	0.78	1.05	1.49	0.78	0.60	1.00	1.12	1.04
35	$(L\tau)_{\eta e}\times10^{10}$, s	1.09	1.62	1.67	1.61	0.60	0.53	0.67	2.58	0.51	0.60	1.02	2.04	
	$(\tau/L)_b\times10^{10}$, s	1.01	1.16	1.67	0.59	0.79	0.98	1.21	1.65	0.81	1.35	1.09	1.75	
	$\tau\times10^{10}$, s	1.05	1.37	1.67	0.97	0.70	0.72	0.90	2.06	0.64	0.90	1.05	1.89	1.16
	L	1.04	1.18	1.00	1.65	0.87	0.73	0.74	1.25	0.79	0.67	0.97	1.08	1.00
40	$(L\tau)_{\eta e}\times10^{10}$, s	0.72	0.78	1.03	0.96	0.43	0.44	0.43	1.40	0.29	0.46	0.75	1.46	
	$(\tau/L)_b\times10^{10}$, s	0.70	1.01	1.54	0.39	0.73	0.62	1.00	0.90	0.54	0.92	0.91	1.31	
	$\tau\times10^{10}$, s	0.71	0.89	1.26	0.61	0.56	0.52	0.66	1.12	0.40	0.65	0.83	1.38	0.80
	L	1.01	0.88	0.82	1.57	0.77	0.84	0.66	1.25	0.73	0.71	0.91	1.06	0.93

6.3.4   PARAMETER B

In accordance with the determination (Eq. 6), the b parameter is a measure of the velocity gradient of hydrodynamic flow created by the working cylinder rotation, influence on characteristic time t_v^* of g action on the shear strain of the m-ball and its rotation movement. Own characteristic time t_m^* of m-ball shear and rotation accordingly to Eq. (16) depends only on ρ, N and T via τ_m.

It is follows from the experimental data (see Table 1) that the b parameter is a function on the all three variables ρ, N and T, but, at that, is increased at T increasing and is decreased at ρ and N increasing. In order to describe these dependences let us previously determine the angular rate ω_m^0 (s^{-1}) of the strained m-ball rotation with the effective radius $R_m L_m$ of the working cylinder by diameter d contracting with the surface:

$$\omega_m^0 = \pi d\omega / R_m L_m \qquad (35)$$

Here π is appeared due to the difference in the dimensionalities of ω_m^0 and ω.

Let us determine the t_v^0 as the reverse one ω_m^0:

$$t_v^0 = R_m L_m / \pi d\omega \qquad (36)$$

Accordingly to Eq. (36) t_v^0 is a time during which the m-ball with the effective radius $R_m L_m$ under the action of working cylinder by diameter d rotation will be rotated on the angle equal to the one radian. Let us note, that the t_m^* was determined by authors [17] also in calculation of the m-ball turning on the same single angle.

Since in our experiments the working cylinder had two rotating surfaces with the diameters d_1 and d_2, the value ω_m^0 was averaged out in accordance with the condition $d = (d_1 + d_2)/2$; so, respectively, the value t_v^0 was averaged out too:

$$t_v^0 = 2R_m L_m / \pi(d_1 + d_2)\omega . \qquad (37)$$

So, t_v^0 is in inverse proportion to ω; therefore, through the constant device it is in inverse proportion to g: $t_v^0 \sim g^{-1}$. However, as it was noted in Chapter 2.2, in m-ball due to the difference in linear rates of the polymeric chains links it is appeared the hydrodynamic interaction which leads to the appearance of the additional to g local averaged upon m-ball velocity gradient of the hydrodynamic flow g_m. This local gradient g_m acts not on the conformational volume of the m-ball but on the monomeric framework of the polymeric chains (the inflexible *Kuhn's* wire model [22]). That is why the endowment of g_m into characteristic time t_v^* depends on the volumetric part φ of the links into the conformational volume of m-ball, i.e., $t_v^* \sim (g + g_m \varphi)^{-1}$.

Therefore, it can be written the following:

$$\frac{t_v^*}{t_v^0} = \frac{g}{g + g_m \phi},$$

(38)

that with taking into account of Eq. (37) leads to the expression

$$t_v^* = \frac{2R_m L_m}{\pi(d_1 + d_2)\omega} \Big/ \left(1 + \frac{g_m}{g}\frac{\rho}{\rho_m}\right).$$

(39)

Combining the Eqs. (16) and (39) into (6) we will obtain

$$b = \frac{7a}{2\pi(d_1 + d_2)} \cdot \frac{L_m}{\tau_m} \Big/ N^{2.4}\left(\frac{\rho}{\rho_0}\right)^2\left(1 + \frac{g_m}{g}\frac{\rho}{\rho_m}\right).$$

(40)

As we can see, here the parameters L_m and τ_m are also inseparable and cannot be found independently one from another. That is why based on the experimental data presented in Table 1 it can be found only the numerical values of the ratio $(\tau_m / L_m)_b$. After the substitution of values $a = 1.86\times10^{-10}$ m, $d_1 = 3.4\times10^{-2}$ m, $d_2 = 3.3\times10^{-2}$ m we have

$$\left(\frac{\tau_m}{L_m}\right)_b = 2.84\cdot10^{-9} \Big/ N^{2.4}\left(\frac{\rho}{\rho_0}\right)^2\left(1 + \frac{g_m}{g}\frac{\rho}{\rho_m}\right) b .$$

(41)

As it was marked in Chapter 2.2, we could not estimate the numerical value of g_m / g due to the smallness of the value η_{pm}^0 lying in the error limits of its measuring. That is why, we will be consider the ratio g_m / g as the fitting parameter starting from the consideration that the concentrated solution for polymeric chains is more ideal than the diluted one and, moreover, the m-ball is less strained than the single polymeric ball. That is why, g_m / g was selected in such a manner that the factor of form L_m was near to the 1. This lead to the value $g_m / g = 25$.

The calculations results of $(\tau_m / L_m)_b$ accordingly to Eq. (41) with the use of experimental values from Table 1 and also the values $g_m / g = 25$ are represented in Table 2. They mean that the $(\tau_m / L_m)_b$ is a visible function on a temperature but not on a ρ and N.

On a basis of the independent estimations of $(\tau_m / L_m)_\eta$ and $(\tau_m / L_m)_b$ it was found the values of τ_m and L_m, which also presented in Table 2.

An analysis of these data shows that with taking into of their estimation error it is discovered the clear dependence of τ_m and L on T, but not on ρ and N. Especially clear temperature dependence is visualized for the values $\tilde{\tau}_m$, obtained via the averaging of τ_m at giving temperature for the all values of ρ and N (Table 2). The temperature dependence of τ_m into the coordinates of the *Arrhenius'* equation is presented on Fig. 6.

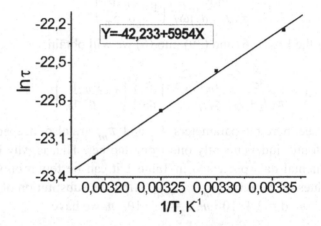

FIGURE 6   Temperature dependence of the average values of the characteristic time τ of the segmental movement of polymeric chain in coordinates of the *Arrhenius* equation

6.4   POLYSTYRENE'S MELT

6.4.1   EXPERIMENTAL DATA

Typical dependencies of the melt viscosity η on the angular rate ω (rotations per second) of the working cone rotation are represented on Figs. 7 and 8.

In order to analyze the experimental curves of $\eta(\omega)$, the Eq. (5) with the same remarks as to the numerical estimations of parameters η_e, η_f and b was used.

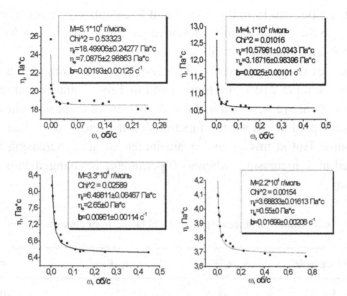

FIGURE 7 Experimental (points) and calculated in accordance with the Eq. (5) (curves) dependencies of the effective viscosity on the velocity of the working cone rotation. T = 210°C.

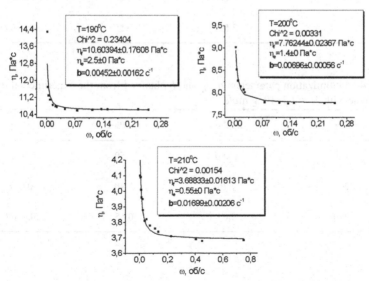

FIGURE 8 Experimental (points) and calculated in accordance with the Eq. (5) (curves) dependencies of the effective viscosity on the velocity of the working cone rotation. $M =$ 2.2×10⁴ g/mo*le*.

As it can be seen from the Figs. 7 and 8, calculated curves of $\eta(\omega)$ accordingly to the Eq. (5) with the founded parameters η_f, η_e and b describe the experimental data very well.

Results of the numerical estimations of η_f, η_e та b on a length of the polymeric chain at 210°C are represented in Table 3 and the temperature dependencies are represented in Table 4. Review of these data shows, that the all three parameters are the functions on the length of a chain and on temperature. But at this, η_e and η_f are increased at N increasing and are decreased at T increasing, whereas b parameter is changed into the opposite way.

TABLE 3 Optimization parameters η_f, η_e and b obtained from the experimental data at T = 210°C.

$M \times 10^{-4}$, g/mole	η_f, Pa·s	η_e, Pa·s	b, s⁻¹
5.1	18.49	7.09	0.0019
4.1	10.58	3.19	0.0025
3.3	6.50	2.65	0.0096
2.2	3.69	0.55	0.0169

TABLE 4 Optimization parameters η_f, η_e and b obtained from the experimental data for polystyrene with $M = 2.2 \times 10^4$ g/*mole*

T, °C	η_f, Pa·s	η_e, Pa·s	b, s⁻¹
190	10.60	2.50	0.0045
200	7.76	1.40	0.007
210	3.69	0.55	0.0169

6.4.2 FRICTIONAL COMPONENT OF THE EFFECTIVE VISCOSITY OF MELT

Results represented in Tables 3 and 4 show that the frictional component of the viscosity η_f very strongly depends on the length of the polymeric chains and on the temperature. The whole spectrum of the dependence of η_f on N and T will be considered as the superposition of the above earlier listed three forms of the motion which make the endowment into the frictional component of the viscosity of melt, *namely* the frictional coefficients of viscosity η_{sm}, η_{pm} and η_{pz} (*see* Chapter 2.2.).

Not the all listed forms of the motion make the essential endowment into η_f, however for the generalization let us start from the taking into account of the all forms. So, the frictional component of the viscosity should be described by the Eq. (19), but for the melt it is necessary to accept that $\eta_s = 0$, and $\varphi = 1$.

$$\eta_f = \eta_{sm} + \eta_{pm} + \eta_{pz}. \tag{42}$$

In a case of the melts η_{pz} is determined by the Eq. (30), but at $\varphi \cong \rho/\rho_m = 1$

$$\eta_{pz} = \eta_{pz}^0 N^2 \left(\frac{\rho}{\rho_0}\right)^{2.5}. \tag{43}$$

Here the coefficient of the proportionality η_{pz}^0 contains the ratio L_m/L_f, however, since the melt is the ideal solution for polymer, that is why it can be assumed that $L_m = 1$.

By substituting of the Eq. (43) into Eq. (40), we will obtain

$$\eta_f = \eta_{sm} + \eta_{pm} + \eta_{pz}^0 N^2 \left(\frac{\rho}{\rho_0}\right)^{2.5} \tag{44}$$

Let's estimate an endowment of the separate components into η_f. The results represented in Table 3 show, that under conditions of our experiments the frictional component of viscosity is changed from the minimal value equal to 3.7 *Pa·s* till the maximal one equal to 18.5 *Pa·s*. The value of the viscosity coefficient η_{sm}, which represents the segmental motion of the polymeric chains and estimated [20] on the basis of η_f for diluted solution of polystyrene in toluene is equal approximately 5×10^{-3} $Pa \cdot s$. Thus, it can be assumed that $\eta_{sm} < \eta_f$ and to neglect by respective component in Eq. (44). Taking into account of this fact, the Eq. (44) let's rewrite as follow:

$$\eta_f = \eta_{pm} + \eta_{pz}^0 N^2 \left(\frac{\rho}{\rho_0}\right)^{2.5}. \tag{45}$$

For the melts $\rho/\rho_0 = const$, that is why the interpretation of the experimental values of η_f as the function of N^2 is represented on Fig. 9. It can be seen from the Fig. 9, that the linear dependence corresponding to the Eq. (45) is observed, and the numerical value of $\eta_{pz}^0 \left(\frac{\rho}{\rho_0}\right)^{2.5}$ was found upon the inclination tangent of the straight line; under the other temperatures this coefficient was found using the experimental data from Table 4. For the estimation of η_{pz}^0 it was assumed that $\rho = 1.05 \cdot 10^6$ g/m^3, ρ_0 was calculated in accordance with Eq. (11) at $M_0 = 104.15$ g/mole, $a = 1.86 \times 10^{-10}$ m.

Temperature dependence of η_{pz}^0 into coordinates of the *Arrhenius* equation is represented on Fig. 10.

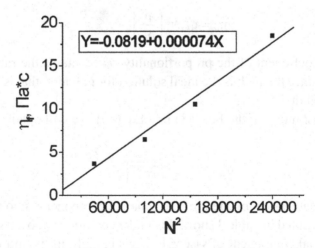

FIGURE 9   Interpretation of the experimental values of η_f in the coordinates of the Eq. (45) at $T = 210°C$.

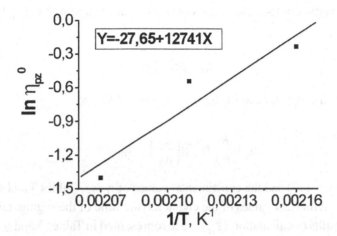

FIGURE 10 Temperature dependence of the numerical estimations of η_{pz}^{0} in the coordinates of the *Arrhenius* equation.

So, an activation energy $E_{pz,}$ of the relative motion of intertwined polymeric chains into polystyrenes' melt consists of 106 ± 35 kJ/mole.

It can be seen from the presented Fig. 9 and from the regression equation, that the values of η_{pz}^{0} are very little and are within the ranges of their estimation error; this cannot give the possibility to estimate the value of g_m/g.

So, as the analysis of the experimental data showed, the main endowment into the frictional component of the effective viscosity of the polystyrenes' melt has the relative motion of the intertwining between themselves into *m*-ball polymeric chains. Exactly this determines the dependence of η_f on the length of a chain ($\eta_f \sim N^2$).

6.4.3 ELASTIC COMPONENT OF THE EFFECTIVE VISCOSITY OF MELT

It can be seen from the Tables 3 and 4, that the elastic component of the viscosity η_e is strongly growing function on a length of a chain N and declining function on temperature T.

The elastic component of the viscosity is described by the Eq. (43), but at $L_m = 1$.

$$\eta_e = \mu t_m^*$$ (46)

Correspondingly, instead of the Eq. (34) we obtained

$$\eta_e = \frac{RT}{M_0} N^{3.4} \rho \left(\frac{\rho}{\rho_0} \right)^{3.5} \tau_m$$ (47)

Using the Eq. (47) and the experimental values of η_e (see Tables 3 and 4) it was found the numerical values of the characteristic time of the segmental motion τ_m. The results of calculation $(\tau_m)_{\eta_e}$ are represented in Tables 5 and 6. Despite the disagreement in numerical estimations, it is observed their dependence on T, but not on the N; this fact is confirmed by the Eq. (47).

TABLE 5 Characteristic times of the segmental motion calculated based on the experimental values of η_e and b ($M = 2.2 \times 10^4$ g/mole).

T, °C	$(\tau_m)_{\eta e} \times 10^{11}$, s	$(\tau_m)_b \times 10^{11}$, s	$\tau_m \times 10^{11}$, s
190	6.86	5.50	6.18
200	3.76	3.58	3.67
210	1.45	1.48	1.47

TABLE 6 Characteristic times of the segmental motion calculated based on the experimental values of η_e та b ($T = 210°C$).

M×10⁻⁴, g/mole	$(\tau_m)_{\eta e} \times 10^{11}$, s	$(\tau_m)_b \times 10^{11}$, s	$\tau_m \times 10^{11}$, s
5.1	1.11	1.70	
4.1	1.05	2.25	
			1.48
3.3	1.84	0.99	
2.2	1.45	1.48	

6.4.4   PARAMETER B

In accordance with the determination (Eq. 6), the b parameter is a mea-sure of the velocity gradient of hydrodynamic flow created by the working cylinder rotation, influence on characteristic time t_v^* of g action on the shear strain of the m-ball and its rotation movement. Own characteristic time t_m^* of m-ball shear and rotation accordingly to Eq. (16) depends only on ρ, N and T via τ_m.

It is follows from the experimental data (see Tables 3 and 4) that the b parameter is a function of N and T, but at this it is increased at T increasing and is decreased at N growth. In order to describe of these dependencies let's previously determine the angular rate $\omega_m^0 (s^{-1})$ of the rotation of m-ball with the effective radius R_m, which contacts with the surface of the working cone with radius r

$$\omega_m^0 = \pi r \omega / R_m \tag{48}$$

Here π is appeared as a result of the different units of the dimension $\omega_m^0, (s^{-1})$ and $\omega, (rot/s)$.

Let's determine the t_v^0 as the reverse one to the ω_m^0

$$t_v^0 = R_m / \pi r \omega \tag{49}$$

In Eq.(49) t_v^0 is a time during which the m-ball with the conformational radius R_m under the action of working cone rotation with radius r will be rotated on the angle equal to the one radian. Let us note, that the t_m^* was determined by authors [17] also in calculation of the m-ball turning on the same single angle.

Thereby, t_v^0 is inversely proportional to ω, so via the constant of the device is inversely proportional to g: $t_v^0 \sim g^{-1}$. However, into m-ball as a result of the difference in the linear rates of the links of polymeric chains under their rotation the hydrodynamic interaction is appeared, which leads to the appearance of the additional to the g local averaged upon m-ball gradient velocity of the hydrodynamic flow g_m. This local gradient g_m acts not on the conformational volume of m-ball, but on the monomeric frame of the polymeric chains (the inflexible Kuhn's wire model [22]). That is why the endowment of g_m into characteristic time t_v^* depends on the volumetric part φ of the links into the conformational volume of m-ball, that is, $t_v^* \sim (g + g_m \varphi)^{-1}$. Into the melt $\varphi = 1$, therefore, it can be written the following:

$$t_v^*/t_v^0 = g/(g + g_m), \tag{50}$$

that leads with taking into account of the Eq. (49), to the expression

$$t_v^* = \frac{R_m}{\pi r \omega} \bigg/ \left(1 + \frac{g_m}{g}\right). \tag{51}$$

By combining of the Eqs. (16) and (51) in (6), we obtained

$$b = \frac{7a}{4\pi r \tau_m} \bigg/ N^{2.4} \left(\frac{\rho}{\rho_0}\right)^2 \left(1 + \frac{g_m}{g}\right). \tag{52}$$

As we can see, using the experimental values of b parameter (see Tables 3 and 4) it can be calculated $(\tau_m)_b$. After the substitution of the values $a = 1.86 \times 10^{-10}$ m, $r = 1.8 \times 10^{-2}$ m, we obtained

$$(\tau_m)_b = 3.78 \cdot 10^{-6} \bigg/ N^{2.4} \left(\frac{\rho}{\rho_0}\right)^2 \left(1 + \frac{g_m}{g}\right) b. \tag{53}$$

Numerical value of the ratio g_m/g was considered as a parameter, which selected in such a way that calculated accordingly to Eq. (53) values of $(\tau_m)_b$ corresponded to the calculated values $(\tau_m)_{\eta_e}$ accordingly to Eq. (47). So, the obtained value of $g_m/g = 39$.

The results of calculations of $(\tau_m)_b$ and $(\tau_m)_{\eta_e}$, are compared in Tables 5 and 6. As the results show, $(\tau_m)_b$ and $(\tau_m)_{\eta_e}$ is visible function on the temperature but not on the N. That is why based on the data of Tables 5 and 6 it were calculated the averaged values of the $\tilde{\tau}_m$ of the characteristic time of segmental motion of the macromolecule.

Temperature dependence of $\tilde{\tau}_m$ into coordinates of the *Arrhenius* equation is represented on Fig. 11.

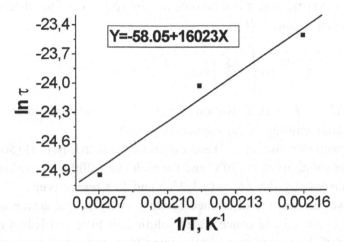

FIGURE 11   Temperature dependence of the averaged values of characteristic time τ of the segmental motion of polymeric chain into coordinates of the *Arrhenius* equation

6.5   CHARACTERISTIC TIME OF THE SEGMENTAL MOTION OF POLYSTYRENE IN SOLUTIONS AND MELT

The presentation of values $\tilde{\tau}_m$ into *Arrhenius'* coordinates equation Figs. 6 and 11 permitted to obtain the expressions for concentrated solutions and melt, respectively:

$$\ln\tilde{\tau}_m = -42.23 + 5950/T, \tag{54}$$

$$\ln\tilde{\tau}_m = -58.05 + 16020/T, \tag{55}$$

For diluted solution of polystyrene in toluene it was early obtained [20]:

$$\ln\tilde{\tau}_f = -44.07 + 6660/T, \tag{56}$$

On a basis of the Eqs. (54)-(56) it was calculated the activation energies of the segmental motion of polystyrene in diluted, concentrated solutions and melt, which consists of 55.4, 49.4 and 133.0 kJ/mole, respectively.

Characteristic time τ can be obtained by equation of the theory of absolute reactions rates [16]:

$$\tau = \frac{2h}{kT}\exp\left\{-\frac{\Delta S^*}{R}\right\}\exp\left\{\frac{\Delta H^*}{RT}\right\} = \tau_0\exp\{E/RT\}, \tag{57}$$

where $\Delta H^* = E$ is an activation energy of the segmental motion; ΔS^* is an activation entropy of the segmental motion.

By comparing the Eq. (57) and experimental data Eqs. (54)-(56), it was found for solutions at $t = 30°C$ and for melt at $t = 200°C$ the values of the activation entropy $\Delta S^*/R = 15.3$, 13.5 and 28.8 respectively.

As we can see, the difference between energies and entropies of activation in diluted and concentrated solutions is little and is in a range of the error of their estimation. At the same time, indicated parameters into melt of the polystyrene is approximately in two times higher. Besides, the growth of the activation entropy does not compensate the activation energy growth; as a result, the characteristic time of the segmental motion into melt is on 2–3 orders higher, than into the solutions (at the extrapolation of τ on general temperature).

Let's compare the values of the activation energies E with the evaporation heats ΔH_{evap} of styrene (-43.94 kJ/mole) and toluene (–37.99 kJ/mole). So, independently on fact, which values of ΔH_{evap} were taken for styrene or toluene, it is observed a general picture: $E_{тm}$, $E_{тf} > \Delta H_{вип}$. It is known Ref. [23], that for the low-molecular liquids, viscosity of which is determined by the Brownian or translational form of the molecules motion, the activation energy of the viscous flow is in 3-4 times less than the evaporation heats. This points on fact, that the segmental motion, which is base of the reptation mechanism of the polymeric chains motion, is determined by their deformation–vibrational freedom degrees.

However, let's mark another circumstance. During the study of the bimolecular chains termination kinetics [24] which is limited by their diffusion, in polymeric matrixes of the dimethacrylate TGM-3 (triethylenglycole dimethacrylate), monomethacrylate GMA (2,3-epoxypropylmethacrylate) and their equimolar mixture TGM-3 : GMA = 1 : 1 in the temperature range 20 ÷ 70°C it were obtained the following values of the activation energies: 122.2, 142.3 and 131.0 kJ/mole. Since the diffusion coefficient of the macroradical is also determined by the characteristic time of the segmental motion, it can be stated that the presented above activation energies of the segmental motion in melt and polymeric matrix are good agreed between themselves. A sharp their difference from the activation energy into solutions points on: 1) a great influence of the solvent as a factor activating the segmental motion of polymeric chain, and 2) on fact, that the dynamic properties of the polymeric chains in melt are very near to their dynamic properties in polymeric matrix.

6.6 DYNAMIC PROPERTIES OF POLYSTYRENE IN SOLUTIONS AND MELT

Dynamical properties of the polymeric chains are determined by characteristic times of their translational motion (t_t^*) and rotation (t_r^*) motions. As it was noted earlier, the characteristic time of the shear strain is also equal to t_r^*. Since the monomeric links connected into a chain, the all of these types of motion are realized exceptionally in accordance with the reptation mechanism that is via the segmental motion with the characteristic time τ_s. That is why, let's analyze and generalize once more the obtained experimental data of the characteristic times of the segmental motion of the chains of polystyrene in solutions and melts, which were estimated based on elastic component of the viscosity η_e and parameter b. Besides, let's add to this analysis the characteristic times of the segmental motion, estimated based on coefficient of the frictional component of viscosity of diluted solution (η_{sm}), concentrated solution and melt (η_p°).

The values τ_s will be used in the sequel for the estimation of the characteristic time of the translation motion t_t^* and of the coefficient of the diffusion D of the polystyrene chains into solutions and melt. Accordingly to the experimental data the temperature dependence τ_s, estimated based on the elastic component of the viscosity η_e and parameter b, is described by the equations:

in diluted solution (temperature range 20–35°C)

$$\ln \tau_s = -44.07 + 6660/T,$$

(58)

in concentrated solution (temperature range 25-40°C)

$$\ln \tau_s = -42.23 + 5950/T \,, \tag{59}$$

in melt (temperature range 190–210°C)

$$\ln \tau_s = -58.05 + 16020/T \tag{60}$$

Let's write also the temperature dependencies of the coefficients of a frictional component of the viscosity:
    in diluted solution

$$\ln \eta_{sm} = -29.04 + 7300/T \,, \tag{61}$$

in concentrated solution

$$\ln \eta_{pz}^\circ = -16.67 + 4800/T \,, \tag{62}$$

in melt

$$\ln \eta_{pz}^\circ = -27.65 + 12740/T \,. \tag{63}$$

Next, let's use the proposed earlier expression for characteristic time of the segmental motion in the following form

$$\ln \tau_s = \ln 2 \frac{h}{kT} - \frac{\Delta S_s}{R} + \frac{E_s}{RT} \,, \tag{64}$$

where $\ln 2 \dfrac{h}{kT} = -28.78$ and -29.22 at $T = 303$ K and $T = 473$ K correspondingly.

Using of these values and comparing Eqs. (64) and (58)-(60), we will obtain the numerical estimations for the activation entropy of the segmental motion $\Delta S_s / R$, which represented in Table 7.

TABLE 7   Characteristic parameters of segmental motion of polystyrene in solutions and melt.

System	E_s, kJ/mole	$\dfrac{\Delta S_s}{R}$	E_{pz}, kJ/mole	$\dfrac{\Delta S_{pz}}{R}$	T = 303 K		T = 473 K	
					τ_s, s	τ_{pz}, s	τ_s, s	τ_{pz}, s
Diluted solutions	55.3	15.3	—	—	$2.5 \cdot 10^{-10}$	—	—	—
Concentrated solutions	49.4	13.5	39.9	6.0	$1.5 \cdot 10^{-10}$	$6.0 \cdot 10^{-9}$	—	$1.9 \cdot 10^{-11}$
Melt	133.1	28.8	105.9	17.0	$5.6 \cdot 10^{10*}$	$1.5 \cdot 10^{-2*}$	$3.1 \cdot 10^{-11}$	$4.0 \cdot 10^{-9}$

Note. *Data found by the extrapolation accordingly to the Eqs. (60) and (71) in the field of the glass-like state of melt

In Table 7 also the activation energies E_s of the segmental motion and the value τ_s at T = 303 K and $T = 473$ K are represented too. Values τ_s at T = 303 K in melt were obtained by the extrapolation of Eq. (60) on given temperature, at which melt is in the solid glass-like state.

It can be seen from the Table 7, that the numerical values both of τ_s, and the thermodynamic characteristics ($\Delta S_s/R$ and E_s) of the segmental motion into diluted and concentrated solutions are differed only within the limits of the experimental error of their estimations. In melt these values are essentially differed. At this, the growth of the activation energy (approximately from 55 kJ/mole till 133 kJ/mole) of the segmental motion is not compensated by the growth of the activation entropy (till $\Delta S_s/R \approx 29$); as a result, the values of τ_s in melt are on two orders greater than in solutions (at $T = 473$ K) and on six orders greater at $T = 303$ K.

Let's assume, that the coefficients of the frictional component of the viscosity of polymeric chains are described by as same general expression [16], as the coefficients of the viscosity of low–molecular solution. At that time it can be written:

$$\eta_{sm} = 3\frac{RT}{V}\tau_{sm}, \tag{65}$$

$$\eta_{pz}^{\circ} = 3\frac{RT}{V}\tau_{pz}, \tag{66}$$

where V is the partial–molar volume of the monomeric link of a chain.

τ_{sm} and τ_{pz} are per sense, the characteristic times of the segmental motion of free polymeric chain into diluted solution and overlapping one with others polymeric chains into the concentrated solution and melt taking into account the all possible gearing effects, correspondingly.

Since the partial–molar volume V of the monomeric link of the polystyrene is unknown, then for the next calculations it can be assumed without a great error to be equal to the molar volume of the monomeric link into the melt:

$$V = \rho/M_0, \tag{67}$$

where $\rho = 1.05 \cdot 10^6$ g/m^3 is a density of the polystyrene melt; $M_0 = 104.15$ g/mole is the molar mass of styrene. Let us write Eqs. (65) and (66) in general form:

$$\ln\eta = \ln 3RT\frac{\rho}{M_0} + \ln\tau. \tag{68}$$

At this $\ln 3RT\frac{\rho}{M_0} = 18.15$ and 18.59 at $T = 303$ K and T = 473 K, correspondingly. Taking into account of this value and comparing the Eqs. (61)–(63) and (68), we will obtain the temperature dependences τ_{sm} and τ_{pz}:
    for diluted solution

$$\ln\tau_{sm} = -47.15 + 7300/T, \tag{69}$$

for concentrated solution

$$\ln\tau_{pz} = -34.82 + 4800/T, \tag{70}$$

for melt

$$\ln \tau_{pz} = -46.24 + 12740/T . \qquad (71)$$

On the basis of two last ones expressions the τ_{pz} have been calculated at T = 303 K and T = 473 K. Taking into account a general Eq. (64) it has been found also the value of the activation entropy $\Delta S_{pz}/R$ (see Table 7).

Comparing the parameters of the Eq. (58) for τ_s and Eq. (69) for τ_{sm}, it can be seen, that the difference between them is adequately kept within the error limits of their estimation. The values of τ_s and τ_{sm} at T = 303 K, equal to 2.5×10^{-10} s and 1.0×10^{-10} s, respectively, prove of this fact. Thus, it can be assumed that $\tau_s \equiv \tau_{sm}$, and that is why the coefficient of the frictional component of the viscosity τ_{sm} of the polymeric chains can be described by as same general Eq. (65) as for the coefficient of the low−molecular solution. The values of τ_{pz} calculated accordingly to the Eqs. (71) and (72) for concentrated solution at T = 303 K and melt at T = 473 K correspondingly (see Table 7), are essentially differed from τ_s: $\tau_{pz} > \tau_s$, approximately on two orders. An analysis of the parameters of the Eqs. (59) (60) and (70) (71) showed that the difference between τ_s and τ_{pz} is caused by two factors, which abhorrent the one of the other: by insignificant decreasing of the activation energy ($E_{pz} < E_s$) that should be decreased the τ_{pz}, and by a sharp decreasing of the activation entropy ($\Delta S_{pz} < \Delta S_s$) that increases of τ_{pz}.

As it was said, the coefficient of the frictional component of viscosity η_{pz} in concentrated solutions and melt caused by the motion of the overlapping between themselves polymeric chains relatively the one of the other and characterizes the efficiency of the all possible gearings. However, the mechanism of this motion is also reputational that is realized via the segmental motion. Correspondingly, between the times τ_s and τ_p the some relationship should be existing. Let's assume the thermodynamical approach for the determination of this relationship as a one among the all possible.

Let's determine the notion "gearing" as the thermodynamical state of a monomeric link of the chain, at which its segmental motion is frozen. This means, that under the relative motion of the intertwining between themselves polymeric chains the reputational mechanism of the transfer at the expense of the segmental motion takes place, but under condition that the part of the monomeric links of a chain is frozen.

Let the ΔG_z° is a standard free energy of the monomeric link transfer from a free state into the frozen one. Then the probability of the frozen states formation or their part should be proportional to the value $\exp\{-\Delta G_z^{\circ}/RT\}$. That is why, if the k_s is a constant rate of the free segmental transfer, and k_{pz} is the rate constant of the frozen segmental transfer, then between themselves the relationship should be existing:

$$k_{pz} = k_s \exp\{-\Delta G_z^{\circ}/RT\}. \tag{72}$$

Then k_{pz}, additionally to k_s, has a free activation energy equal to the standard free defrosting energy of the frozen state.

Since $k_s = \tau_s^{-1}$, $k_{pz} = \tau_{pz}^{-1}$, we obtained

$$\tau_{pz} = \tau_s \exp\{\Delta G_z^{\circ}/RT\}. \tag{73}$$

By assigning

$$\Delta G_z^{\circ} = \Delta H_z^{\circ} - T\Delta S_z^{\circ} \tag{74}$$

and taking into account the experimentally determined ratios $\tau_{pz} > \tau_s$, $E_{pz} < E_s$ and $\Delta S_{pz} < \Delta S_s$, we conclude, that in Eq. (74) $\Delta G_z^{\circ} > 0$, $\Delta H_z^{\circ} < 0$ and $\Delta S_z^{\circ} < 0$, and besides the entropy factor $T\Delta S_z^{\circ}$ should be more upon the absolute value than the enthalpy factor ΔH_z°. These ratios per the physical sense are sufficiently probable. A contact of the links under the gearing can activates a weak exothermal effect $(\Delta H_z^{\circ} < 0)$ at the expense of the intermolecular forces of interaction, and the frosting of the segmental movement activates a sharp decrease of the entropy of monomeric link $\Delta S_z^{\circ} < 0$, but at this $|T\Delta S_z^{\circ}| > |\Delta H_z^{\circ}|$. Let's rewrite the Eq. (73) with taking into account of Eq. (74) in a form

$$\ln\tau_{pz} = \ln\tau_s + \frac{\Delta H_z^{\circ}}{RT} - \frac{\Delta S_z^{\circ}}{R}. \tag{75}$$

Comparing the Eqs. (59), (60), (70) and (72) and taking into account Eq. (75) we obtained:

    for concentrated solution

$$\Delta G_z^\circ = 9.0 \text{ kJ/mole}, \ \Delta H_z^\circ = -9.6 \text{ kJ/mole}, \ \Delta S_z^\circ / R = -7.4,$$

for melt

$$\Delta G_z^\circ = 15.0 \text{ kJ/mole}, \ \Delta H_z^\circ = -31.4 \text{ kJ/mole}, \ \Delta S_z^\circ / R = -1.8.$$

In connection with carried out analysis the next question is appeared: why in the concentrated solutions and melt the gearing effect hasn't an influence on the elastic component of viscosity η_e°, and determined based on this value characteristic time of the segmental motion is τ_s; at the same time, the gearings effect strongly influences on the frictional component of viscosity, on the basis of which the τ_{pz} is estimated. Probably, the answer on this question consists in fact that the elastic component of the viscosity is determined by the characteristic time of the shear, which is equal to the characteristic time of rotation. Accordingly to the superposition principle the rotation motion of the m-ball of the intertwining between themselves polymeric chains can be considered independently on their mutual reloca-tion that is as the rotation with the frozen conformation. As a result, the gearings effects have not an influence on the characteristic time of the rotation motion. Free segmental motion gives a contribution in a frictional component of viscosity, but it is very little and is visible only in the diluted solutions. That is why even a little gearing effect is determining for the frictional component of viscosity in concentrated solutions and melts.

Let's use the obtained numerical values of the characteristic times of the segmental movement τ_s for the estimation of dynamical properties of the polystyrene chains that is their characteristic time of the translational movement t_t^* and coefficient of diffusion D into solutions and melt. Ac-cordingly to Ref. [25], the values t_t^* and D are determined by the expres-sions:

in diluted solutions

$$t_t^* = N^{8/5} \tau_s, \tag{76}$$

$$D = \frac{a^2}{2\tau_s} N^{-3/5} . \tag{77}$$

in concentrated solutions and melt

$$t_t^* = N^{3.4} \left(\frac{\rho}{\rho_0} \right)^{2.5} \tau_s , \tag{78}$$

$$D = \frac{a^2}{2\tau_s} \Big/ \left(\frac{\rho}{\rho_0} \right)^{2.5} N^{2.4} . \tag{79}$$

In order to illustrate the dynamic properties of the polystyrene in solutions and melt in Table 8 are given the numerical estimations of the characteristic times of segmental τ_s and translational t_t^* motions of the polystyrene and diffusion coefficients D. It was assumed for the calculations $a = 1.86 \times 10^{-10}$ m, $N = 10^3$ and $\rho = 0.5 \times 10^6$ g/m³ for concentrated solution and melt correspondingly. As we can see, the characteristic time of the translational motion t_t^* of the polystyrene chains is on 4 and 6 orders higher than the characteristic time of their segmental motion; this is explained by a strong dependence of t_t^* on the length of a chain. The coefficients of diffusion weakly depend on the length of a chain, that is why their values into solutions is on 2–3 order less, than the coefficients of diffusion of low-molecular substances, which are characterized by the order 10^{-9} m²/s.

TABLE 8 Dynamic characteristics of polystyrene in solutions and melt.

System	T = 303 K			T = 473 K		
	τ_s, s	t_t^*, s	D, m²/sc	τ_s, s	t_t^*, s	D, m²/s
Diluted solutions	$2.0 \cdot 10^{-10}$	$1.3 \cdot 10^{-6}$	$1.4 \cdot 10^{-12}$	—	—	—
Concentrated solutions	$2.0 \cdot 10^{-10}$	$2.9 \cdot 10^{-4}$	$1.0 \cdot 10^{-13}$	—	—	—
Melt	$5.0 \cdot 10^{-3*}$	$7.2 \cdot 10^{3*}$	$7.3 \cdot 10^{-22*}$	$3.0 \cdot 10^{-11}$	$4.3 \cdot 10^{-5}$	$1.2 \cdot 10^{-13}$

Note: *Data found by the extrapolation in a field of the glass-like state of melt.

A special attention should be paid into a value of the diffusion coefficient at T = 303 K in a field of the glass-like state of melt $D = 7\times10^{-22}\ m^2/s$. Let's compare of this value D with the diffusion coefficients of the macroradicals in polymeric matrixes TGM-3, TGM-3-GMA and GMA which estimated experimentally [24] based on the kinetics of macroradicals decay, which under the given temperature consist of $10^{-21} \div 10^{-22}\ m^2/s$.

Thus, carried out analysis shows, that the studies of the viscosity of polymeric solutions permits sufficiently accurately to estimate the characteristic times of the segmental and translational movements, on the basis of which the coefficients of diffusion of polymeric chains into solutions can be calculated.

Investigations of a gradient dependence of the effective viscosity of concentrated solutions of polystyrene and its melt permitted to mark its frictional η_f and elastic η_e components and to study of their dependence on a length of a polymeric chain N, on concentration of polymer ρ in solution and on temperature T. It was determined that the main endowment into the frictional component of the viscosity has the relative motion of the intertwined between themselves in m-ball polymeric chains. An efficiency of the all-possible gearings is determined by the ratio of the characteristic times of the rotation motion of intertwined between themselves polymeric chains in m-ball t_m^* and Flory ball t_f^*. This lead to the dependence of the frictional component of viscosity in a form $\eta_f \sim N^2 \rho^{5.5}$ for concentrated solutions and in a form $\eta_f \sim N^2$ for melt, which is agreed with the experimental data.

It was experimentally confirmed the determined earlier theoretical dependence of the elastic component of viscosity for concentrated solutions $\eta_e \sim N^{3.4} \rho^{4.5}$, and for the melt $\eta_e \sim N^{3.4}$, that is lead to the well-known ratio $\eta_e \sim t_m^* \sim N^{3.4}$, which is true, however, only for the elastic component of the viscosity. On a basis of the experimental data of η_e and b it were obtained the numerical values of the characteristic time τ_m of the segmental motion of polymeric chains in concentrated solutions and melt. As the results showed, τ_m doesn't depend on N, but only on temperature. The activation energies and entropies of the segmental motion were found based on the average values of $\tilde{\tau}_m$. In a case of a melt the value of E and $\Delta S^*/R$ is approximately in twice higher than the same values for diluted and concentrated solutions of polystyrene in toluene; that points on a great activation action of the solvent on the segmental motion of the polymeric chain, and also notes the fact that the dynamical properties of the polymeric chains in melt is considerably near to their values in polymeric matrixes, than in the solutions.

An analysis which has been done and also the generalization of obtained experimental data show, that as same as in a case of the low-molecular liquids, an investigation of the viscosity of polymeric solutions permits sufficiently accurately to estimate the characteristic time of the segmental motion on the basis of which the diffusion coefficients of the polymeric chains in solutions and melt can be calculated; in other words, to determine their dynamical characteristics.

KEYWORDS

- **activation energy**
- **effective viscosity**
- **frictional and elastic components of the viscosity**
- **m-ball**
- **segmental motion**

REFERENCES

1. Ferry J. D. Viscoelastic Properties of Polymers / *J. D. Ferry* – *N.Y.: John Wiley and Sons,* 1980, 641 p.
2. De Gennes P. G. Scaling Concepts in Polymer Physics / *P.G. de Gennes* - *Ithaca: Cornell Univ. Press,* 1979, 300 p.
3. Tsvietkov V. N. The Structure of Macromolecules in Solutions / *V. N. Tsvietkov, V. E. Eskin, S. Ya. Frenkel* – *M.: "Nauka",* 1964, 700 p. *(in Russian)*
4. Malkin A. Ya. Rheology: Conceptions, Methods, Applications / *A. Ya. Malkin, A. I. Isayev* - *M.: "Proffesiya",* 2010, 560 p. *(in Russian)*
5. Grassley W. W. The Entanglement Concept in Polymer Rheology / *W. W. Grassley, Adv. Polym. Sci.* - 1974, v.16, 1-8.
6. Eyring H. Viscosity, Plasticity, and Diffusion as Examples of Absolute Reaction Rates / *H. Eyring, J. Chem. Phys.* 1936, v. 4. 283–291.
7. Peterlin A. Gradient Dependence of the Intrinsic Viscosity of Linear Macromolecules / *A. Peterlin, M. Čopic, J. Appl. Phys.* 1956, v. 27. 434-438.

8. Ikeda Ya. On the effective diffusion tensor of a segment in a chain molecule and its application to the nonnewtonian viscosity of polymer solutions / *Ya. Ikeda, J. Phys. Soc. Japan.* – 1957, v. 12. 378–384.

9. Hoffman M. Strukturviskositat und Molekulare Struktur von Fadenmolekulen / *M. Hoffman, R. Rother, Macromol. Chem.* 1964, v. 80. 95–111.

10. Leonov A. I. Theory of Tiksotropy / *A. I. Leonov, G. V. Vynogradov, Reports of the Academy of Sciences of USSR.* 1964, v. 155. № 2. 406-409.

11. Williams M. C. Concentrated Polymer Solutions: Part II. Dependence of Viscosity and Relaxation Time on Concentration and Molecular Weight / *M. C. Williams, A. I. Ch. E. Journal.* 1967, v. 13, № 3, 534-539.

12. Bueche F. Viscosity of Polymers in Concentrated Solution / *F. Bueche*, J. Chem. Phys. 1956, Vol. 25, 599–605.

13. Edvards S. F. The Effect of Entanglements of Diffusion in a Polymer Melt / *S. F. Edvards, J. W. Grant, Journ. Phys.* 1973, v. 46. 1169–1186.

14. De Gennes P. G. Reptation of a Polymer Chain in the Presence of Fixed Obstacles / *P. G de Gennes, J. Chem. Phys.* 1971, v. 55, 572-579.

15. Medvedevskikh Yu. G. Gradient Dependence of the Viscosity for Polymeric Solutions and Melts / *Yu. G. Medvedevskikh, A. R. Kytsya, L. I. Bazylyak, G.E Zaikov, Conformation of Macromolecules. Thermodynamic and Kinetic Demonstrations – N. Y.: Nova Sci. Publishing,* 2007, 145–157.

16. Medvedevskikh Yu. G. Phenomenological Coefficients of the Viscosity of Low-Molecular Simple Liquids and Solutions / *Medvedevskikh Yu. G., Khavunko O. Yu., Collection Book: Shevchenko' Scientific Society Reports,* 2011, v. 28, 70–83 *(in Ukrainian).*

17. Medvedevskikh Yu. G. Viscosity of Polymer Solutions and Melts / *Yu. G. Medvedevskikh, Conformation of Macromolecules Thermodynamic and Kinetic Demonstrations – N. Y.: Nova Sci. Publishing,* 2007, 125–143.

18. Medvedevskikh Yu. G. Statistics of Linear Polymer Chains in the Self-Avoiding Walks Model / *Yu. G. Medvedevskikh, Condensed Matter Physics.* 2001, vol. 2. № 26, 209–218.

19. Medvedevskikh Yu. G. Conformation and Deformation of Linear Macromolecules in Dilute Ideal Solution in the Self-Avoiding Random Walks Statistics / *Yu. G. Medvedevskikh, Journ. Appl. Polym. Sci,* 2008, v.109. № 4.

20. Medvedevskikh Yu. Frictional and Elastic Components of the Viscosity of Polysterene–Toluene Diluted Solutions / *Yu. Medvedevskikh, O. Khavunko, Chemistry and Chemical Technology.* 2011, v. 5, № 3, 291–302.

21. Medvedevskikh Yu. G. Conformation of Linear Macromolecules in the Real Diluted Solution / *Yu. G. Medvedevskikh., L. I. Bazylyak, A. R. Kytsya, Conformation of Macromolecules Thermodynamic and Kinetic Demonstrations – N. Y.: Nova Sci. Publishing,* 2007, 35–53.

22. Kuhn H. Effects of Hampered Draining of Solvent on the Translatory and Rotatory Motion of Statistically Coiled Long-Chain Molecules in Solution. Part II. Rotatory Motion, Viscosity, and Flow Birefringence / *H. Kuhn, W. Kuhn, J. Polymer. Sci.* 1952, v. 9, 1–33.

23. Tobolsky A. V. Viscoelastic Properties of Monodisperse Polystirene / *A. V. Tobolsky, J. J. Aklonis, G. Akovali, J. Chem. Phys.* 1965, v. 42, № 2, 723-728.

24. Medvedevskikh Yu. G. Kinetics of Bimolecular Radicals Decay in Different Polymeric Matrixes / *Yu. G. Medvedevskikh, A. R. Kytsya, O. S. Holdak, G. I. Khovanets, L. I. Bazylyak, G. E. Zaikov, Conformation of Macromolecules Thermodynamic and Kinetic Demonstrations – N. Y.: Nova Sci. Publishing,* 2007, 139–209.
25. Medvedevskikh Yu. G. Diffusion Coefficient of Macromolecules into Solutions and Melts / *Yu. G. Medvedevskikh, Conformation of Macromolecules. Thermodynamic and Kinetic Demonstrations – N. Y.: Nova Sci. Publishing,* 2007, 107–123.

CHAPTER 7

POLYANILINE / NANO-TIO$_2$-S HYBRIDE COMPOSITES

M. M. YATSYSHYN, A. S. KUN'KO, and
O. V. RESHETNYAK PETRONIS

CONTENTS

SUMMARY

The polyaniline (PAn)/nano-TiO_2-S nanocomposites were synthesized in the 0.5 M HCl aqueous solutions by oxidation of aniline in the presence of nanoparticles of mineral component (S-doped TiO_2). The physico-chemical properties of produced samples were studied with use of X-ray diffraction phase analysis, FTIR and EDX spectroscopy, SEM microscopy, thermogravimetric and electrochemical methods. It was determined that the morphology of synthesized samples changes from aggregative (3D net structure) to thin polymeric layers on the surface of TiO_2-S nanoparticles with increasing of the mineral component content in the composite. In the last case a polyaniline is characterized by sufficiently ordering of macromolecules in the result of interfacial interaction, namely H-bonding between polymeric macromolecules and surface of TiO_2-S nanoparticles. Besides, it was determined, that the increasing of content of the mineral component leads to the increasing of the thermal stability of the PAn/nano-TiO_2-S composite in comparison with the individual polyaniline, but to the decreasing of electric conductivity of samples.

7.1   INTRODUCTION

Composites of conducting polymer (*CP*) and inorganic oxides are very attractive and perspective materials for different branches of science, *namely* for chemistry, physics, electronics, photonics etc. due to synergetic effect which arises under the integration of the properties of oxide and *CP* [1, 2]. In the present time the titanium (*IV*) oxide (rutile and anatase crystalline modifications) is one among widely used inorganic oxides, which is applied for the chemical synthesis of such nanocomposites [3-7].

Nanosized TiO_2 (anatase) is nontoxic, chemically inert and inexpensive material. It is actively studied and widely used in the electronics, photonics [1] and especially in the photocatalysis [8]. In recent years TiO_2 has been studied because it is also known as a photocatalyst [9], which accelerates the formation of the hydroxyl radicals under light. Hydroxyl radicals are powerful oxidizing agents that can disinfect and deodorize air, water, and surfaces in environmental-decontamination applications [10–

13]. The features of nanosized TiO$_2$ induced with the presence of oxygen vacancies that leads to the occurrence of specific chemical and physical defects. However, the high-energy gap width of the nano-TiO$_2$ (3.2 eV) renders impossible its effective operation under irradiation by sunlight. The decreasing of TiO$_2$ energy gap width requires the TiO$_2$ modification by different photosensitizers. For this purpose sulfur is frequently used [14−17], because their energy gap width is 2.9 eV [18].

Other widely used in last time photosensitizer for nano-TiO$_2$ is the polyaniline (*PAn*) with energy gap width 2.8 eV [19]. Polyaniline is organic metal with high conductivity in the doping form ($\sim 10^2$ $S\times$cm^{-1}) [20], good stability, nontoxic, inexpensive and multifunctional polymer and also has the wide perspectives for usage [21−23]. These margins of polyaniline use develop considerably if the possibility of change of main polyaniline forms in the result of effect of different factors will be taken into consideration [21−23].

The combination of the properties of nano-TiO$_2$ and polyaniline enables to solve successfully the problems of the chemistry, physics and electronics. Specific electronic structures of the nano-TiO$_2$ (as the *n*-type semiconductor) and polyaniline (as the electron's conductor in majority of the cases and as a p-type semiconductor under certain conditions) give the possibility to design the systems for different applications. For example, today such materials are equipped in the photocatalytic conversions of the different pollutants especially [14]. The modification of the surface of *TiO$_2$* particles by polyanilines layers raises the catalytic activity of titanium (IV) oxide [5, 19]. Composite materials, which have integrated properties of *S*-doped nano-TiO$_2$ and polyaniline layers can be effective in the photocatalytic processes especially.

PAn/TiO$_2$ composites are produced mainly during the aniline chemical oxidation in situ by different oxidants, for example $(NH_4)_2S_2O_8$ or $Na_2S_2O_8$, in the aqueous solutions of inorganic (HCl, H_2SO_4 etc.) or organic (for example salicylic, toluene sulfonic etc.) acids in the presence of produced previously micro or nanosized *TiO$_2$* particles [13, 22−26]. Polyaniline layers are applied on the surfaces of *TiO$_2$* both without previous modification of oxide particles surface [13] and after such modification, for example by g-amino-propyl-triethoxysilane [27]. Synthesis of the hybrid *PAn/nano-TiO$_2$* composites can be carried out also by one-step method [28−29], exactly in the micellar solutions of surfactants [30].

The composites particles on the basis of polyaniline and micro or nano-sized TiO_2 can be characterized by different structure, namely core-shell type [5], microrods [31], microspheres [1–3, 5, 13, 23. 32], nanowires [28], nanonets [33], which content the individual TiO_2 particles and their aggregates, and also the polymeric films with different thickness and morphology on the surface of the oxide nanoparticles [24]. The shape and size of the PAn/nano-TiO_2 composite particles are determined by synthesis conditions of the nanosized TiO_2 and composite on its base respectively.

The combination of polyaniline, nano-TiO_2 and other components [34–35] permits to produce the composites with improved physico-chemical properties. Such nanocomposite materials are studied actively and are employed in the different branches of engineering and technics [31, 36], as cathodic materials in the chemical power sources [37], in the electronics [7, 38], chemo- and biosensors [39–40], and also as the components of corrosion protection coverages [41] or protective shades of different assignments [42].

It can be expected that the PAn/nano-TiO_2-S composite also has the attractive features and also arrives at a wide use in the modern life. Therefore the main aim of our work is the synthesis and study of physico-chemical properties of such composite materials.

7.2 EXPERIMENTAL

7.2.1 MATERIALS

Aniline (Aldrich, 99.5%) was distilled under the reduced pressure of 4 *Torr* and stored under argon. Other used in the work reagents were analytical grade and used without the additional purification. The solutions of chloride acid were prepared with use of the standard titers. For preparation of the all solutions the twice distilled water as used.

The nanoparticles of sulfur-doped TiO_2 (anatase) were synthesized by solid state method. In this case the initial metha-titanic acid H_2TiO_3 was dehydrated in the presence of sulfur (~1.5 mass percent) during its heating in the muffle furnace under the 500°C over the time of 2 h. The synthesized sample was cooled then in the exiccator to the room temperature. An analysis of *SEM*-images of the synthesized TiO$_2$-*S* particles indicated that they are characterized by an aggregative structure with the size of aggregates, which form the spherical individual nanoparticles by ~20 *нм* size [*33*].

The samples of PAn and PAn/nano-TiO$_2$-*S* composites were produced by aniline oxidative polycondensation in the presence of sodium peroxydisulfate as an oxidant in the 0.5 *M* aqueous *HCl* solution under the temperature 2 ± 0.5°C. Aniline : *Na$_2$S$_2$O$_8$* molar ratio was 1 : 1.1. During the synthesis the 0.01 mole of aniline dissolved in the 50 cm^3 of 0.5 *M* aqueous *HCl* solution, introduced the batch of *S*-doped titanium (IV) oxide (0.25−2.0 *g*), exposure this mixture by ultrasonic machining for 10 min for the disaggregation of nano-TiO$_2$-*S* particles and stood over the time of 1 h. In this solution under the mixing adds (by the ~0.5 cm^3 portions) 50 cm^3 solution of *Na$_2$S$_2$O$_8$* (0.011 *mole*) in the cooled to the temperature 2 ± 0.5°C 0.5 M HCl solution and mixes this reaction mixture further °C over the time of 1 h at the same temperature. The reaction mixture further held at room temperature for 24 h, filtered, washed off by 250 cm^3 of distilled water to neutral value of pH and dried in the vacuum drier under the 60 ± 1°C and under pressure 0.9 ± 0.05 kg·*cm^{-2}*. Preparation conditions of synthesized samples of PAn and composites shown on Table 1.

The reaction of the oxidative condensation of aniline in the *HCL* aqueous solutions can be presented by following scheme [*43*]:

TABLE 1 Preparation conditions of synthesized samples of PAn and PAn/nano-TiO$_2$-*S* composites.

	Sample	PAn	C2	C1	C0.75	C0.5	C0.25
Reagents	Aniline / *mol*	0.01	0.01	0.01	0.01	0.01	0.01
	HCl / M	0.5	0.5	0.5	0.5	0.5	0.5
	Na$_2$S$_2$O$_8$ / mol	0.011	0.011	0.011	0.011	0.011	0.011
	nano-TiO$_2$-*S* / g	–	2.0	1.0	0.75	0.50	0.25

The synthesized product was a powder by green color that indicates about production of the emeraldine form of polyaniline, exactly emeraldine salt PAn-*HCl*. The produced samples of PAn/nano-TiO$_2$-*S* composites had more light tinctures; their intensities were decreased with the increasing of the TiO$_2$ content in the initial reaction mixture.

7.2.2 INSTRUMENTAL

The physico-chemical properties of the produced samples of PAn, nano-*TiO$_2$-S* and its composites were characterized with the use of the following experimental methods and instrumental supplies. X-ray phase analysis were carried out with use of *DRON*-3 diffractometer (*Cu Ka* radiation, l = 1.54 Å). *FTIR* spectra of synthesized samples were registered on the *NICOLET IS* 10 spectrometer (samples were pressed into a pellet together with *KBr*).

Paulic-Paulic-Erdei Q-1500 *D* derivatograph was used during studies of the thermal stability of synthesized samples. Measurements were carried out in the 20-900°*C* temperature interval (dynamic mode with a heating speed of 10°K·min^{-1}) in an air atmosphere and with use of corundum pots. The sample's weight was 200 *mg*, reference substance - *Al$_2$O$_3$*.

Scanning electron microscopy *(SEM)* images and results of the energy dispersive *X*-ray *(EDX)* spectroscopy were obtained on the *JEOL JSM-6400* microscope. The parallel beam of electrons with energy 3 kV was used for *SEM*-image acquisition of the samples surface, while under the qualitative and quantitative determination of elements in the samples the energy of electrons *(EHT)* was scanned in the 0−10 kV interval. In the first case the distance from electron source to samples surface was 3–4 mm, while during *EDX* studies - 10 mm.

The cyclic voltammetry *(CVA)* measurements were carried out using a computer-controlled potentiostat / galvanostat 50–1 (Ukraine). Powders of synthesized polymer or composites were pressed into pellets with thickness ~2 mm and diameter 10 mm under the pressure 150 *atm·cm^{-2}* and temperature 20°C over the time of 5 min and were used as the working electrodes. Platinum sheet was used as an counter electrode. The potential was scanned in the (−200 , (+600) mV interval with a 50 mV·s^{-1} scanning rate. All experimental potential values are referred to the saturated silver / silver chloride electrode.

Conductivity of produced samples was measured under the 20°C with use of sandwich-type cell. In this case the pellets, which were produced analogously as working electrodes for voltammetry, were used during the conductivity measurements.

7.3 RESULTS AND DISCUSSION

7.3.1 STRUCTURAL STUDIES OF PAN/NANO-TiO$_2$-S COMPOSITES

The results of the X-ray phase analysis of the synthesized samples of nano-TiO$_2$-S, PAn and PAn/nano-TiO$_2$-S composites are shown on Fig. 1. There are several sharp peaks, namely at $2q$ = 25.3, 38.1, 48.2, 54,9, 62.9, etc., on the diffractogram of nano-TiO$_2$-S sample (Fig. 1, *curve 1*), which correspond to the crystalline structure of anatase [*13, 28*]. The diffraction peaks at $2q$ equal 9.6, 14.9, 21.4, 25.6, 27.6, 29.6 and 34.5° on the diffractogramm of polyaniline (Fig. 1, *curve 7*) are typical for the high ordered structures of doped polyaniline PAn-*HCl* [*30, 44*] and corresponds to reflections of crystal planes (001) (011) (100) (110) (111) and (020) for PAn pseudoorthorhombic structure [*45*]. The presence of these seven peaks on the background of wide halo (Fig. 1, *curve 7*) indicated that the synthesized sample of individual polyaniline has a mixed amorphous-crystalline structure [*44*]. The patterns exhibit sharp diffraction peaks at $2q$ = 25.6° and 21.4°, which can be ascribed to periodicity parallel and periodicity perpendicular to the polymer chain, respectively [*46*]. The peak at $2q$=25.3° is most intensive on the *TiO$_2$* diffractogramm (Fig. 1, *curve 1*). It is agreed practically with PAn characteristic peak at 25,6°, which disappears during polymerization deposition of polyaniline on the surface of nano-*TiO$_2$-S* particles, but accrues peak at $2q$=25.6° (Fig. 1, *curve 2-6*). This fact indicates that majority of polyaniline molecules in the composites are deposited on the surface of *TiO$_2$-S* nanoparticles whose crystalline structure to bring influence on the crystallinity of PAn [*28*]. The observed high crystallinity and orientation of polyaniline macromolecules on the surface of nano-TiO_2 particles is promissory because such high-ordered systems show higher electric conductivity [44].

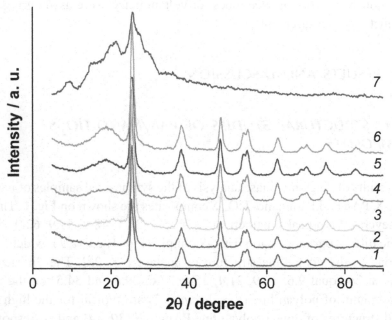

FIGURE 1   *X*-ray diffractogramms of nano-TiO_2-*S* (*1*), PAn (*7*) and PAn/nano-TiO_2-*S* composites: **C2** (*2*), **C1** (*3*), **C0.75** (*4*); **C0.5** (*5*), and **C0.25** (*6*) samples.

There are *FTIR* spectra of synthesized samples of individual PAn, nano-TiO_2-*S* and their composites on the Fig. 2. The form of *FTIR* spectrum *7* in the 400–4000 cm^{-1} interval and set observed characteristic bands, namely at 3400–2880, 1560, 1470, 1290, 1110 and 793 cm^{-1}, corresponds to polyaniline [28]. In particular, the peak at 3400–2880 cm^{-1} corresponds to so-called "H-band" which is a visualization of H-bonding imine- (-NH-) and protonated imine-group (-NH$^+$-) polyaniline macromolecules between itself [46]. Two characteristic bands at the 1560 and 1470 cm^{-1} are connected with vibrations of quinoid and benzenoid rings and they are characteristic features of the polyaniline. Intensive peaks at 1290, 1110 and 793 cm^{-1}, corresponds to the emeraldine salt and indicated on the high doping degree of PAn [46].

The slight shift of characteristic bands at 1560 and 1470 cm^{-1} observed for the samples of PAn/nano-TiO$_2$-S composites (Fig. 2, *spectra* 2–5). It is a result of the interfacial interaction which takes place between PAn macromolecules and TiO$_2$-S nanoparticles. N-H...O and N...H−O H-bonding are realized in this case between imino-groups polymeric links and surface O= and HO- group of TiO$_2$, respectively [46]. The increasing of intensity of the wide band at 3 500−3 800 cm^{-1} with maximum at 3 700 cm^{-1} (Fig. 2, *spectra* 2–5) is confirmation of this conclusion [3]. At once, the increasing of TiO$_2$-S content in the composite leads to intensive absorption of *IR* radiation in the 200–800 cm^{-1} interval

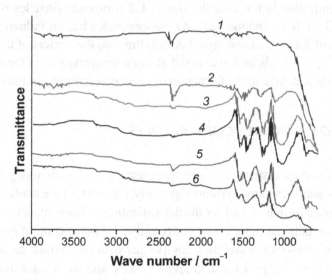

FIGURE 2 *FTIR* spectra of the of nano-TiO$_2$-S (*1*), PAn (*6*) and PAn/nano-TiO$_2$-S composites: C1 (*2*), C0.75 (*3*); C0.5 (*4*), and C0.25 (*5*) samples.

7.3.2 MORPHOLOGY OF NANO-TiO$_2$-S COMPOSITES

SEM images of the initial nano-TiO$_2$-S and synthesized composites (**C1** and **C2**) are shown on the Fig. 3. The difference between images under the low TiO$_2$-S content is absent practically, while the morphology of samples under the high contents of mineral component is different. The analysis of *SEM-*

image of synthesized TiO_2-S particles (Fig. 3, a–c) indicated that they are characterized by aggregative structure with size of aggregates from 200 to 800 нм. Analogous morphology is typical also for the synthesized PAn/nano-TiO_2-S composites, however the size of aggregates is slightly smaller - from 50 нм and they are looser (Fig. 3, d-i). The images which were obtained under the magnitude ×100.000 (Fig. 3, b–c) confirms that aggregates of TiO_2-S are formed by individual nanoparticles with size ~20 нм [33]. Polyaniline in the **C1** composite (Fig. 3, e, f) forms $3D$ net structure, which annexes individual TiO_2 particles and their fine aggregates. Under the higher TiO_2 content (Fig. 3, h) polymer deposits as a thin layer on the surface of individual mineral nanoparticles that favors disaggregation of the formed particles of PAn/nano-TiO_2-S composite. In this case the size of **C2** composite particles is ~25–30 nm (Fig. 3, i). It is obvious that TiO_2 nanoparticles has an influence on the formation of the polyaniline layer because they are the centers of the polymerization [3] that leads, as it was noted above during analysis of the results of X-ray diffraction data, to the formation of more ordered polyaniline deposits.

7.3.3 EDX SPECTROSCOPY RESULTS

Spectra of energy dispersive X-ray spectroscopy in the chemistry of CP, in the case of nanostructured systems especially, are used for the confirmation of the composite formation and for the determination of the content of inorganic oxides or metallic particles components in them. EDX spectra of synthesized samples of nano-TiO_2-S and their composites with polyaniline are shown in the Fig. 4. Intensive peaks at 4.55 and 4.95 keV and weak-intensive peak at 0.45 keV corresponds to titanium atoms while the peak at 0.52 keV to the atoms of oxygen (Fig. 4, a). The presence of sulfur as a doping agent in the TiO_2 confirms weak intensive peaks at 2.31 and 2.47 keV, respectively (Fig. 4a). There are additional weak-intensive peaks at 0.27 and 2.62 keV in the EDX spectra of PAn/nano-TiO_2-S composites (Fig. 4, b–c), in comparison with spectrum of individual TiO_2, which corresponds to carbon and chlorine atoms (doping agent of PAn in the emeraldine form) respectively. Signal from nitrogen atoms (at 0.40 eV [47]) is absent in the measured spectra of composites. However, as it is known, it appears in the polyaniline EDX spectra not always [48–49], which is connected with special state of these atoms in the polyaniline macromolecules probably.

TABLE 2 Results of **energy dispersive** *X*-**ray spectroscopy** of synthesized samples (respective spectra are shown on Fig. 4, *a–c*).

Ele-ment	Samples					
	nano-TiO$_2$-S		C1		C2	
	Mass fraction / percent	Atomic fraction / percent	Mass fraction / percent	Atomic fraction / percent	Mass fraction / percent	Atomic fraction / percent
C K	–	–	34.31	49.36	18.93	32.74
O K	44.53	68.86	36.96	39.91	36.62	47.57
S K	1.31	0.71	1.33	0.72	1.19	0.77
Cl K	–	–	1.03	0.50	1.03	0.60
Ti K	53.52	30.43	26.37	9.51	42.23	18.32
Σ	100.00	100.00	100.00	100.00	100.00	100.00

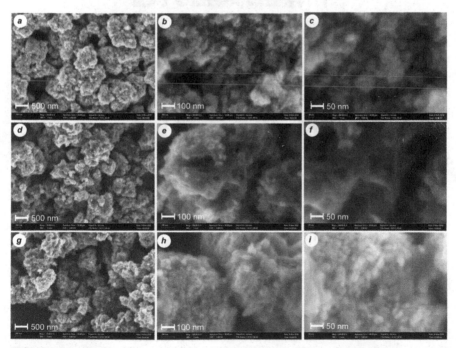

FIGURE 3 *SEM*-images of individual nano-TiO$_2$-*S* (a–c), **C1** (*d–f*) and **C2** (*g–i*) PAn/nano-TiO$_2$-*S* composites with ×20,000 (*a, d, g*), ×100,000 (*b, e, h*) and ×200,000 (*c, f, i*) magnification, respectively.

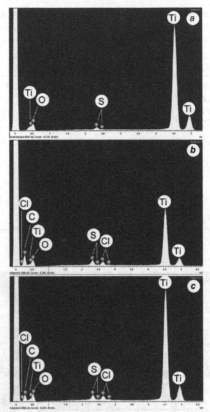

FIGURE 4   *EDX* spectra of individual nano-TiO$_2$-*S* (a), **C1** (*b*) and **C2** (*c*) PAn/nano-TiO$_2$-*S* composites

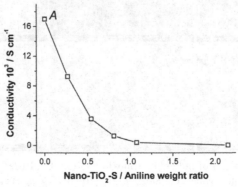

FIGURE 5   Dependence of the conductivity of synthesized samples of PAn/nano-TiO$_2$-*S* composites from the weight ratio nano-TiO$_2$-*S* : PAn in the reaction mixture during their synthesis.

The quantitative results of *EDX* spectra analysis (Table 2) permitted to conclude about the contents of samples. The juxtaposing of the results of *X*-ray phase analysis (Fig. 1) and *EDX* spectra (Table 2) indicates that PAn/nano-TiO$_2$-*S* composites contains nanosized-TiO$_2$-S, polyaniline and Cl$^-$-anions as its doping agent. Decrease of the carbon content in the composite **C2** in comparison with **C1** even distribution is evidence of more even distribution of polymer in the composite in this case, exactly the formation thin polyaniline layer on the surface of mineral particles.

7.3.4 CONDUCTIVITY AND CYCLIC VOLTAMMETRY STUDIES

Results of conductivity studies of synthesized samples are shown on Fig. 5. As it expected, the highest conductivity observes the polyaniline sample - 17.0×10^{-3} *S* cm^{-1} (Fig. 5, *dot A*). The introduction and further increasing of semiconductor TiO$_2$-*S* nanoparticles content in the composites leads to the exponential decreasing of the conductivity of samples.

The samples of PAn, **C0.25** and **C0.5** composites were tested as electrode materials. The studies were carried out in the 0.5 *M* sulfate acid aqueous solution under the potential scanning in the (−200)-(+600) mV interval. The cyclic volyammogramms for the first five cycles of potential scanning (Fig. 6) demonstrates a good conductivity and reversibility of the produced samples. The higher conductivity, as it was noted above, characterizes the polyaniline sample. In this case, there is hysteresis loop on the cyclic voltammogamm (Fig. 6, *a*), which is decreased with the introduction of mineral component in mineral matrix (Fig. 6, *b*) and for the **C0.5** composite the voltammogramm corresponds to Ohm's law practically (Fig. 6, *c*). The values of cathodic and anodic currents are commensurate for the polyaniline (Fig. 6, a), while the higher values of anodic current in comparison with cathodic observes for the composite's electrodes (Fig. 6, *b−c*), that can be explained by doping-dedoping processes of polyaniline component of composite during electrode cyclic polarization or less overvoltage of hydrogen evolution in the presence of *TiO$_2$*.

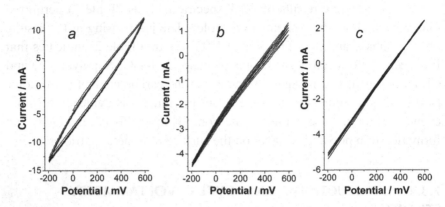

FIGURE 6 Cyclic voltammogramms (5 cycles of potential scanning) of the individual PAn (a), **C0.25** (b) and **C0.5** (c) PAn/nano-TiO$_2$-S composites in the 0.5 M H$_2$SO$_4$ aqueous solution.

7.3.5 THERMOGRAVIMETRIC STUDIES

The results of thermogravimetric studies of produced samples, namely individual nano-TiO$_2$-S and polyaniline, and its composite are shown on the Figs. 7-8 and in Table 3. The character of obtained curves of mass loss Δm (TG), differential curves of mass loss (DTG) and differential thermal analysis (DTA) are different considerably (Fig. 8).

The sample of nano-TiO$_2$-S is characterized by high thermal stability in the 20–900°C temperature interval, where the total mass loss (Δm$_{900}$) is » 5% (Fig. 7, curve 7; Fig. 8, a, curve 1), while the total mass loss for the PAn sample is reached to » 95% (Fig. 7, curve 1; Fig. 8, b, curve 1). Therefore, the main mass loss in the composites is connected with the thermal decomposition of polymeric component and values of total mass loss increases with increasing of PAn content in the samples (Fig. 8, curves 2-6). In the same time, the increasing of TiO$_2$-S nanoparticles in the samples leads to the decreasing of thermodestruction rate of polyaniline component. It indicates the lower grade of TG curves in the 230–600°C temperature's interval. Besides, temperature of the termination of Pan component thermal destruction in the composites samples is shifted in the side of lower temperatures, whereas the start temperature of composites destruction (liberation of doping agent) is shifted in the side of the higher values in comparison with the individual polyaniline. Such shifts indicate on the strong interaction between the particles of titanium (IV) oxide and the polyaniline molecules in the composites [50].

Three stages can be marks out on the thermogravimetric curves (Fig. 8, *a–c, curves 1*) during the samples heating (results for **C1** sample are presented as an example of thermal characteristics of synthesized composites). Results of the quantitative analysis of these data are shown in Table 3. The first stage of mass loss (section *A-B* on the *curves 1*, Fig. 8) is connected with desorption of physically fixed water, while the second stage (interval *B-C*) – with liberation of the water, which *H*-bonded with PAn macromolecules, doping agent *(HCl)* and hydration water of doping Cl⁻-anions [51–55]. The limited sulfur burning from *TiO$_2$-S* particles takes place for sample of individual mineral component (Fig. 8, *a*) in the in the *C-D* interval on the *TG* curves, while oxidative destruction of polymer observes mainly in the same interval for the Pan and composites samples. The mass losses of the individual polyaniline sample for the noted intervals averages $\approx$ 11.5% (T $\approx$ 50–132°C), $\approx$ 2.5% (132–250°C) and $\approx$ 73.5% (250–900°C), respectively.

Analysis of the results presented in Table 3 indicates that the increasing of TiO$_2$-S nanoparticles content in the composites leads to the certain decreasing of mass losses on the all marked out sections of curves *1* (Fig. 8, *b–c*), that confirms above noted explanations about nature of mass losses during oxidative destruction of Pan and composites samples. Temperature T_1 (*see* Fig. 8) of first maximum on *DTG* curves decreases from 98°C to 66°C from PAn to nano-TiO$_2$-S individual samples. In the same time the value of temperatures T_2 of second maximum increases from 285°C to 350°C under the introduction of mineral component in polymer matrix and with its further increasing of its content. The several discrepancies of T_2 values can be stipulated by inorganic oxide influence on the processes on the second stage of thermal destruction of samples. In particular, the hydrated state of the surface of TiO$_2$-S nanoparticles and, as a result, the participation of terminated *HO*-groups in formation *H*-bonds with polymer macromolecules must be taken into consideration. Therefore the value T_2 = 129°C (Table 3) for the individual TiO$_2$-S samples can be connected with the process of dehydration of the surface of nanoparticles.

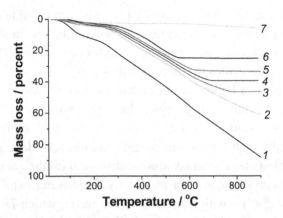

FIGURE 7 Thermogravimetric *(TG)* curves of the polyaniline *(1)*, nano-*TiO₂*-*S* *(7)* and their composites: **C0.25** *(2)*, **C0.5** *(3)*, **C0.75** *(4)*, **C1** *(5)* and **C2** *(6)*.

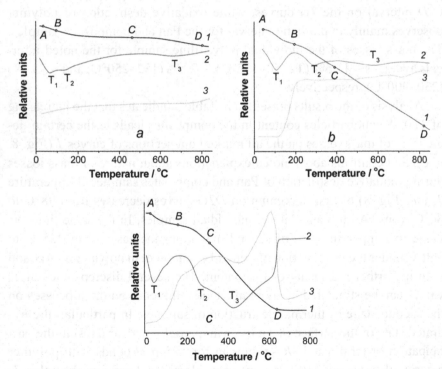

FIGURE 8 *TG (1)*, *DTA (1)* and *DTG (1)* curves for the synthesized samples of nano-TiO₂-*S* *(a)*; *PAn* *(b)* and **C1** composite *(c)*.

TABLE 3 Results of the analysis of *TG* and *DTG* curves of studied samples.

Samples	Mass loss of samples on the section of TG curve / percent			Total mass loss Δm_{900} / percent	Temperatures that corresponds to maxima of mass losses / (± 1.0) °C		
	A - B*	B - C*	C – D*		T_1^*	T_2^*	T_3^*
PAn	11.5	2.5	73.5	95.2	98	285	554
C0.25	4.4	2.3	54.7	61.4	95	300	500
C0.5	4.2	2.7	39.2	46.7	85	325	485
C0.75	4.0	2.9	32.3	42.4	82	342	479
C1	3.4	3.2	28.0	34.6	81	290	470
C2	2.8	3.7	19.5	26.0	78	350	483
TiO$_2$-S	2.1	0.6	2.25	5.0	66	129	670

*See Fig. 8 (a–c) and comments in the text.

The temperature T_3 is decreased with the increasing of TiO_2 content in composites, but for the **C0.5 - C2** samples it is constant practically. Such character of T_3 change, in our opinion, is connected with the changes in morphology of produced samples - from aggregative (3*D* net structure in the **C0.25** and **C0.5** samples) to thin polyaniline layers on the surface of mineral nanoparticles (for the **C0.5-C2** samples), that assured fuller oxidative destruction of polymeric component of composite under lower temperatures. However, the more detailed analysis of *DTG* curves of produced samples with high content of nano-TiO_2-*S* indicated that the third temperature interval (*C-D*, see Fig. 8) contains several successive steps of polymer destruction. Besides main step of polyaniline oxidative destruction, it are the steps of oligomeric fraction of polyaniline liberation, supplementary finish liberation of doping agent, final carbonization of intermediate chemicals [56] and limited sulfur burning from TiO_2-*S* particles (with T_3 = 670 °C, see Table 3), that confirms the complicacy of the destruction of composites samples.

7.4 CONCLUSIONS

The PAn/nano-TiO$_2$-S composites were synthesized in the 0.5 $M HCl$ aqueous solutions by oxidative oxidation of aniline in the presence of nanoparticles of mineral component with size ~ 20 nm. The results of X-ray diffraction phase analysis, $FTIR$ and EDX spectroscopy, thermogravimetric studies indicates that polyaniline is deposited on the surface of mineral component particles and is characterized by sufficiently ordering of macromolecules. The change of samples morphology from aggregative ($3D$ net structure) to thin polymeric layers on the surface of TiO_2-S nanoparticles with increasing of mineral component content was detected under the analysis of SEM images.

The formation of composite, but not mechanical mixture of components, confirms the results of EDX and $FTIR$ spectroscopy. Moreover, the shift of characteristic bands in the $FTIR$ spectra of composites samples confirms the presence of strong interfacial interaction between polymeric macromolecules and surface of TiO_2-S nanoparticles in the result of H-bonding.

It was determined, that the increasing of content of the mineral component leads to the increasing of the thermal stability of the PAn/nano-TiO$_2$-S composite in comparison with individual polyaniline, but to the decreasing of electric conductivity of samples.

It is expected that the synthesized composites with the high contents of TiO_2-S, when the thin polyaniline layer deposited on the mineral nanoparticles, will be effective sensitizers in the processes of the photooxidation in consequence of the synergetic effect occurrence under the integration of the properties of S-doped TiO_2 and polyaniline.

KEYWORDS

- **conductivity**
- **crystallinity**
- **interfacial interaction**
- **nanocomposite**
- **polyaniline**
- **surface morphology**
- **thermal stability**
- **titanium (IV) oxide**

REFERENCES

1. G. Korotcenkov, *"Mater. Sci. Engineer. B"*, **139**, 1–23 (2007).
2. M. Wan. Conducting Polymers with Micro or Nanometer Structure, *"Tsinghua University Press"*, Beijing; *"Springer-Verlag GmbH"*, Berlin-Heidelberg, 312 p. (2010).
3. J.-C. Xu, W.-M. Liu, H.-L. Li, *"Mater. Sci. Engineer. C"*, **25**, 444-447 (2005).
4. N. Parvatikar, S. Jain, S. Khashim, et al., *"Sens. Actuators B"*, **114**, 599-603 (2006).
5. X. Wang, S. Tang, C. Zhou, J. Liu, W. Feng, *"Synth. Met."*, **159**, 1865–1869 (2009).
6. L. Shi, X. Wang, L. Lu, et al., *"Synth. Met."*,**159**, 2525–2529 (2009).
7. C. Bian, A. Yu, H. Wu, *"Electrochem. Commun."*, **11**, 266–269 (2009).
8. L. Sánchez, J. Peral, X. Doménech, *"Electrochim. Acta"*, **42**, 1872–1882 (1997).
9. Su S.-J., Kuramoto N., *"Synth. Met."*, **114**, 147–153 (2000).
10. H. Zhang, R. L. Zong, J. C. Zhao, Y. F. Zhu, *"Environ. Sci. Technol."*, **42**, 3803–3807 (2008).
11. R. K. Mohammad, H. Y. Jeong, S. L. Mu, T. L. Kwon, *"React. Funct. Polym."*, **68**, 1371–1376 (2008).
12. L. Zhang, P. Liu, Z. Su, *"Polym. Degrad. Stabil."*, **91**, 2213–2219 (2006).
13. Zhang L., Wan M., Wei Y., *"Synth. Met."*, **151**, 1–5 (2005).
14. C. M. The, A. R. Mohamed, *"J. Alloys Compd."*, **509**, 1648–1660 (2011).
15. A. Koca, M. Sahin, *"Int. J. Hydrogen Energy"*, **27**, 363–367 (2002).
16. T. Umebayashi, T. Yamaki, H. Itoh, K. Asai, *"Appl. Phys. Lett."*, **81**, 454–456 (2002).
17. H. Tian, J. Ma, K. Li, J. Li, *"Ceram. Int."*, **35**, 1289–1292 (2009).
18. T. Umebayashi, T. Yamaki, S. Yamamoto, A. Miyashita, S. Tanaka, T. Sumita, K. Asai, *"J. Appl. Phys."*, **93**, 5156–5160 (2003).
19. Li X., Wang D., Cheng G., Luo Q., An J., Wang Y., *"Appl. Catalysis B: Environment"*, **81**, 267–273 (2008).
20. S.-C. Kim, D. Kim, J. Lee, Y. Wang, K. Yang, J. Kumar, F.F. Bruno, L.A. Samuelson, *"Macromol. Rapid Commun."*, **28**,1356–1360 (2007).

21. A. Malinauskas, *"Polymer"*, **42**, 3957–3972 (2001).

22. Ye. Koval'chuk, M. Yatsyshyn, N. Dumanchuk, *"Proc. Shevchenko Sci. Soc. Chem. Biochem."*, **21**, 108–122 (2008).

23. Nanostructured conducting polymers / A. Eftekhari (Ed.), *"John Wiley & Sons, Ltd."*, Chichester, UK, 800 p. (2010).

24. H. Tai, Y. Jiang, G. Xie, J. Yu, X. Chen., *"Sens. Actuators B"*, **125**, 644–650 (2007).

25. B.-S. Kim, K-T. Lee, P.-H. Huh, D.-H. Lee, N.-J. Jo, J.-O. Lee, *"Synth. Met."*, **159**, 1369–1372 (2009).

26. M. Oh, S. Kim, *"Electrochim. Acta"*, **78**, 279–285 (2012).

27. F.-Y. Chuang, S.-M. Yang, *"Synth. Met."*, **152**, 361–364 (2005).

28. C. Bian, Y. Yu, G. Xue, *"Appl. Polym. Sci."*, **104**, 21–26 (2007).

29. K. U. Savitha, H. G. Prabu, *"Mater. Chem. Phys."*, **130**, 275–279 (2011).

30. J. Yang, Y. Ding, J. Zhang, *"Mater. Chem. Phys."*, **112**, 322–324 (2008).

31. M. R. Karim, J. H. Yeum, M. S. Lee, K. T. Lim, *"React. Funct. Polym."*, **68**, 1371–1376 (2008).

32. T.-C. Mo, H.-W. Wang, S.-Y. Chen, Y.-C. Yeh, *"Ceram. Int."*, **34**, 1767–1771 (2008).

33. M. M. Yatsyshyn, E. P. Koval'chuk, A. S. Kun'ko, V. S. Kun'ko, Kh. Besaga, *"Material Science of Nanostructures"*, No. **1**, 81-86 (2011).

34. H. Zhou, C. Zhang, X. Wang, H. Li, Z. Du., *"Synth. Met."*, **161**, 2199–2205 (2011).

35. H. Çetin, B. Boyarbay, A. Akkaya, A. Uygun, E. Ayyildiz., *"Synth. Met."*, **161**, 2384–2389 (2011).

36. S. Ameen, M. S. Akhtar, G.-S. Kim, Y. S. Kim, O.-B. Yang, H.-S. Shin, *"J. Alloys Compd"*, **487**, 382–386 (2009).

37. K. Gurunathan, D. P. Amalnerkar, D. C. Trivedi, *"Mater. Lett."*, **57**, 1642–1648 (2003).

38. Y. Li, J. Hagen, D. Haarer, *"Synth. Met."*, **94**, 273–277 (1998).

39. J. Zheng, G. Li, X. Ma, Y. Wang, G. Wu, Y. Cheng, *"Sens. Actuators B"*, **133**, 374–380 (2008).

40. H. Tai, Y. Jiang, G. Xie, J. Yu, X. Chen, Z.Ying //*"Sens. Actuators B"*, **129**, 319–326 (2008).

41. S. Radhakrishnan, C. R. Siju, D. Mahanta, S. Patil, G. Madras, *"Electrochim. Acta"*, **54**, 1249–1254 (2009).

42. S. W. Phang, M. Tadokoro, J. Watanabe, N. Kuramoto, *"Curr. Appl. Phys."*, **8**, 391–394 (2008).

43. I. Sapurina, J. Stejskal, *"Polym. Int."*, **57**, 1295–1325 (2008).

44. A. Rahy, D. J. Yang, *"Mater. Lett."*, **62**, 4311–4314 (2008).

45. J. P. Pouget, M. E. Jòzefowicz, A. J. Epstein, et al., *"Macromolecules"*, **24**, 779–789 (1991).

46. X. Li, *"Electrochim. Acta"*, **54**, 5634–5639 (2009).

47. M.R. Karim, H.W. Lee, I.W. Cheong, S.M. Park, W. Oh, J.H. Yeum, *"Polym. Composite."* **31**, 83-88 (2010).

48. X.-X. Liu, L. Zhang, Y.-B. Li, L.-J. Bian, Y.-Q. Huo, Z. Su., *"Polymer Bull."*, **57**, 825–832 (2006).

49. M. Karthikeyan, K. K. Satheeshkumar, K. P. Elango, *"J. Hazardous Mater."*, **163**, 1026–1032 (2009).

50. J. Deng, C. L. He, Y. Peng, J. Wang, X. Long, P. Li, A. S. C. Chan, *"Synth. Met."*, **139**, 295–301 (2003).

51. K. Pielichowski, *"Solid State Ionics"*, **104**, 123–132 (1997).
52. D. Tsocheva, T. Zlatkov, L. Terlemezyan, *"J. Therm. Anal."*, **53**, 895–904 (1998).
53. P.C. Rodrigues, G. P. de Souza, J. D. D. M. Neto, L. Akcelrud, *"Polymer"*, **43**, 5493–5499 (2002).
54. I. S. Lee, J. Y. Lee, J. H. Sung, H. J. Choi, *"Synth. Met."*, **152**, 173–176 (2005)..
55. N. Salahuddin, M. M. Ayad, M. Ali, *"J. Appl. Polymer Sci."*, **107**, 1981–1989 (2008).
56. S. Bhadra, S. Chattopadhyay, N. K. Singha, D. Khastgir, *"J. Appl. Polymer Sci."*, **108**, 57–64 (2008).

CHAPTER 8

SIMULATION OF CORROSIVE DISSOLUTION OF PT BINARY NANO-CLUSTER IN ACID ENVIRONMENT OF POLYMER ELECTROLYTE MEMBRANE (PEM) FUEL CELLS

S. A. KORNIY, V. I. KOPYLETS', and V. I. POKHMURSKIY

CONTENTS

SUMMARY

A corrosive dissolution model has been proposed for corrosive dissolution binary nanocluster surface of the formula Pt_nX_m (where X – transition metals Cr, Fe, Co, Ni, Ru) of the core-shell structure in acid environment of low temperature fuel cells, which content molecules and ions of H_2O, Cl^-, OH^-, H_3O^+. This model is based on calculations of adsorption behaviors of the environment component interaction with surfaces and activation barriers of platinum atom dissolution by means of quantum-chemical methods PM6 and DFT. Physical-chemical behaviors were established for structural and energy degradation of surfaces of platinum binary nanoclusters with core-shell structure $Pt_{42}X_{13}$ of different composition under the influence of H_2O, Cl^-, OH^-, H_3O+. It was shown that transition metals Cr, Fe, Co, Ni, Ru, which form the core of our binary nanoclusters influence sufficiently on their adsorption behaviors and stability to corrosive dissolution, For example, $Pt_{42}Co_{13}$ is resistant to chlorine hydrated ions influence $Cl^-(H_2O)$ because of much lower adsorption heat (27.5 kJ/mole) as compared with pure platinum nanocluster (195.8 kJ/mole), which should be explained by enhanced electron density in four-fold positions of the surface. It was discovered, that formation of a complex $[Pt(OH)(H_2O)_4]^+$ is possible on the surface of binary nanocluster surface in the environment of water molecules, oxygen atoms and hydroxonium ions, the activation energy of its departure from the surface being the highest for nanoclusters with cobalt and ruthenium in the row $Pt_{55} < Pt_{42}Co_{13} < Pt_{42}Ru_{13}$. It was also shown, that local adsorption of chlorine ions on the hydrated surface of platinum nanoclusters may induce blocking of catalytic centers of their surface due to high adsorption ability of such ions or due to dissolution of platinum atoms with the formation of complexes $[H_2O(PtCl)]^{n+}$, with lower activation energies of their departure from surface in comparison with pure Pt_{55} (more than in twice) with the exclusion of $Pt_{42}Co_{13}$, where an opposite effect was fixed (increase of activation energy in 60%).

8.1 INTRODUCTION

In last decades for purposes of effective nanocatalyst preparation in polymer electrolyte membrane *(PEM)* fuel cells binary nanoclusters of platinum have been considered as perspective materials. Their common formula is Pt_nMe_m (where Me – transition metals Cr, Fe, Co, Ni, Ru) and sizes range from 1 to 50 nm. Experimental [1–3] and theoretical [4–5] studies confirm the enhanced growth of their catalytic activity in comparison with pure metals. Besides, such nanoclusters cost much less. However, in practice effective exploitation time of such catalysts can be sufficiently reduced due to their degradation in the result of oxidation or corrosive dissolution in acid environment [6]. In other words, despite the high catalytic properties of binary nanoclusters, their stability and corrosive resistance is not available for a long period of work. The stability of catalytic layers is determined by the nature of catalysts, environment character, current densities and others. The available experiments show that the main factors in destruction of functionality of PEM fuel cells are as follows: catalytic layer degradation [7–8], corrosion (dissolution) of carbon sublayers [9] and degradation of membrane diffusive layers [10]. Among them degradation of catalytic particles is a main reason of catalytic activity reduction due to next factors: corrosion or dissolution of nanoclusters, their poisoning or passivation during surface segregation of harmful particles, which slow main reaction, surface layer composition changes after their interaction with solution and reagent, agglomeration of nanoclusters with reduction of their effective surfaces.

A big amount of experimental studies of stability of many component systems Pt_nMe_m (where Me – transition metals Cr, Fe, Co, Ni, Ru) indicates about the formation of nanoclusters with core-shell structures [11–13], where mechanisms of the processes (including corrosive) with the formation of such structures are described. Firstly this is a surface segregation during the process of multicomponent nanocluster preparation [14]. Due to such segregation nanocluster surface becomes enriched by one of the components, especially by platinum with the reduction of surface energy in segregated binary nanocluster [15]. In the process of corrosive influence (in model conditions or in tests of fuel cells) a prevailing dissolution of one component from basic metal Me and surface enrichment by platinum with the formation of a "core-shell" system.

Thus, in the work [16] corrosive stability of catalysts $PtCo/C$ and Pt/C were compared in solution of 0.5 M H_2SO_4. Temperature has been increased periodically up to 60°C and tests were made in presence of 10%

H_2O_2 or air bubbles. Experimental data showed that amount of solved platinum for $PtCo/C$ (the alloy exists in two phases: Pt_3Co i $PtCo$) is lower than for Pt/C and it never growth with corrosive activity of solution. However, in this paper exclusively corrosive test results without comparison of catalytic activity of prepared binary alloys are presented. Other works [17–18] also confirm stability of potential of platinum catalysts with cobalt during their long exploitation, compared with pure Pt. Better properties of catalysts $PtCo$ are conjugated with their atomic structure (for example, shorter interatomic distance Pt-Pt) and other electronic behavior of surface, which cause decrease of adsorbed ion OH^- number and catalytic active sites growth up on the surface.

However, the available experimental data about degradation or dissolution of nanocatalysts need theoretical explanation on atomic and molecular levels in different models, with their specific reactions on nanoparticles and size effects in nanosystems. Unfortunately, modern theoretical approaches in simulations often lead to indefinite information about properties of nanoclusters, because of omitting quantum characters of interactions in such systems.

Realistic models must account the structure of surface, shape and dimension of nanoclusters, as well as their chemical composition, which sufficiently change both their catalytic activity and corrosive resistance. It should be noted, that there are no theories of nanocluster stability improvements, developed up to date.

Studies of kinetics and mechanisms of metal cluster ionization in water solutions are of great importance for deepening of theory of catalyst and corrosion. In this respect the development of micromodels is very urgent, including elementary acts of this constituent process – development of structural, electron and energy changes on the way of ionized atom delay from the surface of metals with the formation of hydrated complex is of great importance.

The aim of the present work is to develop a model of corrosive dissolution of Pt binary nanocluster Pt_nMe_m (Me – Cr, Fe, Co, Ni, Ru) in working environment of low temperature fuel cells on the basis of quantum-chemical methods application and deduction of physico-chemical peculiarities of Pt (with different structure and elementary content) surface destruction under the influence of H_2O, Cl^-, OH^-, H_3O^+.

8.2. METHODOLOGY OF COMPUTATION

In present work processes of dissolutions are considered for stable platinum nanoclusters and binary nanocluster $Pt_n Me_m$ (Me is the transition metals, namely *Cr, Fe, Co, Ni, Ru*) with core-shell structure and $n = 42$, $m = 13$ atoms, as well as their surface interaction with particles of environment, i.e., H_2O, O_2, OH^-, H_3O^+, Cl^-. So, we need examine reaction ability of binary nanoclusters in such environment, including prediction of the nanosystem properties. Evidently, to solve such task one must possess data about potentials of all particle interaction in a chosen system.

For successful theoretical simulation in this case we must pick up a correct model in physical aspect with realistic interaction potentials. As previous results in our works [19–20] and other authors' [21–22] indicate, full description of interaction nanocluster – environment requires a complex approximation. Thus we used an approximation of common application of semiempirical quantum-chemical and functional density methods.

Theoretical consideration of platinum nanoclusters originates from the consumption that atoms in an nanocluster should be located at positions with minimal energy in the force field of surroundings. For this purpose we set definite positions of atoms with an approximate geometry and run optimization by molecular dynamics method to obtain minimal energy in dependence on internal coordinates.

References show the various statements regarding the reality of used methods by means of molecular mechanics with atom-atom potentials [23] to study chemical composition and structure of metallic clusters. Sometimes sufficient discrepancies exit between optimizes structures and experimental or ab initio calculated values.

Any structure may be constructed for binary nanoclusters of platinum with transition metals of fourth or fifth row, while in accordance with material science solid solutions of substitution are easy formed. Our previous calculations [24] showed, that nanocluster forms, which are close to cubo-octahedral with 13 and 55 atoms, are the least nanoclusters with maximal internal energy, are the most stable ones and familiar to relatively quick theoretical study by *DFT* methods on *PC*. Besides, such nanoclusters are able to exist in experiments, for instance during *LASER* evaporating of platinum and deposition of particles with mass spectrometric calibration on sublayers. Cubo-octahedral structure allows different changes both composition and locations atoms, including the most real, which are close to experimental values even for much greater clusters with magic numbers of atoms.

Our cubo-octahedral structures were of the "core-shell" type with inner core, formed by second component atoms – transition metals, while the shell in just one atomic layer was constructed by platinum – active catalyst of surface processes. Such structures in our own calculations [25] and others [26–27] are optimal in catalytic sense, because they cause effective way of surface reactions for oxygen reduction. On the other side such nanoclusters possess stability in aggressive acid environments, which lead to electrochemical corrosion of materials of catalysts.

It is known [28] that among random atomic formations with $f.c.c.$ structure of Pt cubo-octaedron Pt_{55} is the most closed parked. It is constructed by eight $f.c.c.$ lattices of platinum with account of symmetry and maximal surface. Two factors support such choice: 1) such size corresponds to four coordinate spheres of the $f.c.c.$ structures; 2) nanoclusters with such sizes are the most stable both in quantum-chemical studies [29] and mass spectrometric measurements. The nearest neighbor number is the greatest one in comparison with similar fragments of crystal lattice. This is the reason of stability.

Nanocluster models were formed on the basis of Pt lattice with the principle of closed atom packing. Binary nanoclusters were built by substitution of Pt atoms by Co, Cr, Fe, Ni, Ru. Valence states in our DFT method were accepted as depicted in Table 1. In common case during the formation of 55-atomic cubo-octahedral Pt nanocluster from a metal specimen a sufficient curvature take place for cluster geometry. Thus, equilibrium relaxed geometry was obtained molecular mechanics approximation as implemented in $HyperChem$ 8.0 [30] software. After such calculations stable in energy 55 atomic clusters were fixed with distances, shown in Table 2. Calculated interatomic distances are less than app. 10%, if compared with volume and satisfy other authors data [31].

TABLE 1 Geometry and electron properties of transition metals in present study of binary nanoclusters.

Type of atom	Type of lattice	Lattice pa-rameter, Å	Ionization potential, eV	Bind energy, eV/atom	Valence state
Pt	f.c.c	3.92	8.96	5.85	$5d^9 6s^1$
Co	h.c.p	3.56	7.86	4.38	$3d^7 4s^2$
Cr	b.c.c	2.88	6.76	4.10	$3d^5 4s^1$
Fe	b.c.c	3.64	7.90	4.29	$3d^6 4s^2$
Ni	f.c.c	3.52	7.63	4.43	$3d^8 4s^2$

We have calculated binary nanoclusters $Pt_{42}Co_{13}$ with core-shell structure, where Pt is the only type of atom, while Co builds the core of nanocluster. For $Pt_{42}Co_{13}$ with core-shell interatomic distances $Pt - Pt$ on its surface are slightly increased up to 2.95 A (Table 2). Bulk distances $Co - Co$ are also enhanced and equal to a middle value of distances for crystal bulk in Pt and Co. Surface Pt atoms are separated from bulk Co atoms in the core-shell model to 2.61 A. The fact is explained by differences in atomic radii of Pt and Co ($r_p/r_{Co} = 1.112$).

TABLE 2 Geometry of Platinum nanoclusters.

	r_{Pt-Pt}, Å	r_{Pt-Co}, Å	r_{Co-Co}, Å
Crystal platinum (experiment)	2.77	–	–
Crystal cobalt (experiment)	–	–	2.50
Nanocluster Pt_{55}, cubo-octaedron	2.52	–	–
Nanocluster $Pt_{37}Co_{18}$, cubo-octaedron, "core shell"	2.95	2.61	2.64

Nanocluster Pt_{55}

Nanocluster $Pt_{42}Co_{13}$
(core–shell)

FIGURE 1 Cubo-octahedral structure of Pt nanoclusters.

For geometry optimization of nanoclusters we used firstly semiempirical method *PM6* [*32*] of the *MOPAC* 2009 [*33*] code. The approximation *PM6*, as literature confirms [*34*], is widely used in calculations of reactivity for systems of transition metals. Although some mistakes in electronic structure studies, geometries are obtained with acceptable accurance and with shot computer time resources.

Electronic structure of optimized nanoclusters was calculated by the program *StoBe* 2009 [*35*] in *DFT* methods with *GGA* approximation [*36*] for exchange-correlation functional B88-*LYP* [*37–38*] and basis set double-ζ and valence polarization *DZVP* [*39*]. Basis sets 6–31G** were accepted for oxygen, chlorine and hydrogen. Relativistic effects were not accounted. The following valence states were taken for elements: $O(2p^4 2 s^2)$ $Cl(3p^5 3 s^2)$, $Co(3d^7 4 s^2)$ and $Pt(5d^9 6 s^1)$.

For the gradient correction the *B*88 approximation for the exchange energy functional and the *LYP* approximation for the correlation functional were employed. The *B*88 exchange functional has the correct asymptotic behavior for the energy density. It reduces the error in the exchange energy by almost two orders of magnitude relative to the *LSDA* result, and thus represents a substantial improvement for a simple functional form containing only one adjustable parameter. In order to increase the computational efficiency, the internal atomic layers are kept frozen for every atom of nanoclusters, since the internal electrons do not contribute significantly to the bonding. A double-zeta basis set of Slater type orbitals was used for valence electrons which include an additional set of functions for each element, playing the role of polarization functions. The frozen core orbitals include up to 4*d* and 3*d* orbitals for *Pt* and *Co* atoms, respectively on the central part of the nanocluster that represent the metallic surface. For the other surrounding atoms, the frozen core orbitals include up to 4*f* and 3*d* orbitals for *Pt* and *Co* atoms, respectively.

Adsorption heat *H* was calculated as a difference between total energy of the system nanocluster-adsorbate and a summa of total energies nanocluster + adsorbat: $H = W_{Pt(PtCo)X} - (W_{Pt(PtCo)} + W_x)$. For simulation of nanocluster dissolution in water solution next assumptions were done. Platinum ionization process may occur in three consequent stages: 1) Platinum dissolution: $Pt \rightarrow Pt^{2+} + 2e^-$; 2) platinum oxide formation: $Pt + H_2O \rightarrow PtO + 2H^+ + 2e^-$; 3) platinum oxide dissolution $PtO + 2H^+ \rightarrow Pt^{2+} + H_2O$.

According to the theory [40], during ionization middle intermediate valence ions concentration is very low. So, we accept that the only product of metal ionization is the ion in the highest valence.

Under the formation of double layer in interface Pt – water solution of $NaCl$ a model was proposed that for positive charged surfaces an ionic parts of double layer are constructed of halohenide with effective radius 1.33 A. Then effective thickness of the $Helmholz$ layer d is equal to ion radii and the potential drops within this thickness.

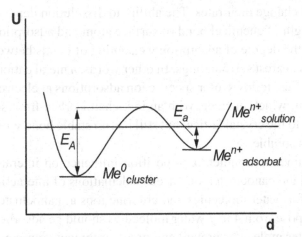

FIGURE 2   A potential energy curve for adiabatic reaction $Pt^0 = Pt^{n+} + ne$: E_A – activation energy of platinum atom departure into solution; E_a – energy of desorption of hydrated platinum ion, yielding during anodic process.

For a correct description of energy changes we estimated adsorption energy of water, strength of chemical bonds for water solution ions with nanocluster surface, changes in surface atom bonds with the rest of cluster and the surface complex metal – chlorine ion was simulated with solvent effects.

Accounting high values of electron state densities on the $Fermi$ level, we accepted that the ionization process for the surface of binary nanoclusters has adiabatic character ($Born$-$Oppenheimer$ approximation) and it is described by a smooth potential curve of transition from the initial state (atom in nanocluster) to final (ion in Me^{n+} solution) [41] (Fig. 2).

8.3   CALCULATION RESULTS

8.3.1   ADSORPTION BEHAVIORS OF WATER MOLECULE INTERACTION WITH THE SURFACE OF PLATINUM BINARY NANOCLUSTERS

According to the Ref. [42], acid anions, such as Cl^-, ClO_4^-, HSO_4^- participate directly in the process platinum nanocluster dissolution in electrolytes and change their rates. The ability to dissolution depends not only on the strength of chemical bond of surface atom and adsorption complex but also on the degree of adsorption weakening of bonds between surface atom and its nearest surroundings. In other words, a metal cation departure from the surface follows after specific ion adsorptions at electrodes. Metal surface atom, which interacts with an ion, chemisorbed from solution, do not belongs to the crystal lattice, but still has no stable bond with complex, more or less soluble.

To examine environment composition influence on interatomic bond alteration in the nanocluster, one needs calculations of interaction through adsorption for water molecules and chlorine ions at nanocluster surfaces. For this purpose from 4 to 7 water molecules should be adsorbed and considered. Water molecules prevail among environment components, where catalytic layers work in fuel cells. So, in cluster approximation adsorption model was constructed for these molecules on the surface (100) of pure nanocluster Pt_{55} and binary nanoclusters $Pt_{42}Me_{13}$. The surfaces (100) possess the higher surface energy and have lower close-parking as compared with (111). Thus the surface (100) shows structural and energy changes first of all, and tends to more intensive dissolution. Calculation has been performed for adsorption heat of water on nanocluster surface, as well as its influence on surface charge alteration on individual atoms and interatomic bond values for $Pt - Pt$ and $Pt - Me$. At the initial stage water molecule is adsorbed on nanoclusters with oxygen atom, oriented to the surface (100) in atop positions (Fig. 3a). In this case addition positive charges on platinum surface atoms and energy decrease for interatomic bonds were observed for all situations (Table 3).

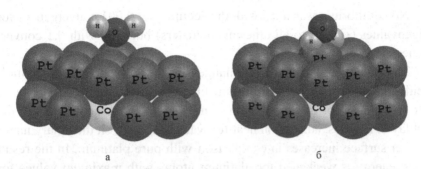

a   б

FIGURE 3 Adsorption position of water molecule on the surface of binary nanocluster $Pt_{42}X_{13}$ before *(a)* and after geometry optimization *(b)*.

TABLE 3 Calculated behaviors of water molecule adsorption on the surface (100) of binary nanoclusters $Pt_{42}X_{13}$: Δq_1 – charge increase on Pt surface atoms, Δq_2 – charge increase on the sublayer atoms (second component), ΔE – bond energy decrease for surface atoms $Pt – Pt$, H – water molecule adsorption heat, r – adsorption distances between hydrogen atoms in water molecule and nanocluster surface.

	Formula of nanocluster					
	Pt_{55}	$Pt_{42}Cr_{13}$	$Pt_{42}Fe_{13}$	$Pt_{42}Co_{13}$	$Pt_{42}Ni_{13}$	$Pt_{42}Ru_{13}$
$\Delta q_1/\Delta q_2$, %	8.5	10.4/5.3	10.8/5.4	12.5/10.7	18.1/12.8	9.3/3.6
ΔE, %	8.5	10.5	12.4	18.6	21.5	10.3
H, kJ/mole	56.25	75.6	98.8	130.5	143.2	65.84
r, nm	0.028	0.026	—	0.015	-0.023	0.056

Charges and energies change most significantly for nickel clusters while slightly for ruthenium clusters. Charge alteration on the second component atoms is below platinum atoms, indicating the local character of water interaction with the surface. Table 3 includes also distances between hydrogen atoms of H_2O and surface of Pt atoms. After geometry optimization (Fig. 3*b*) H atoms locate close to metal surface, almost at the middle of $Pt – Pt$ bond at bridge site and even below the surface for nanocluster $Pt_{42}Ni_{13}$

No equilibrium was fixed with the accuracy of 0.001 convergences for eigenvalues (similar to all other nanoclusters) but only with 0.1 convergences in the case of $Pt_{42}Fe_{13}$.

The results confirm surface changes by the influence of water molecule. Partly during its equilibrium chemisorption the electron charge grows up both for Pt surface atoms and second component atoms in the sublayer. Besides, adsorption heat for water molecule on the binary nanocluster surface increases in comparison with pure platinum. In the result bond energy is weakened for platinum atoms with maximum values for $Pt_{42}Co_{13}$ and $Pt_{42}Ni_{13}$. Thus, $Pt_{42}Co_{13}$ and $Pt_{42}Ni_{13}$ are the most hydrophilic among the structures, here considered.

8.3.2. PLATINUM NANOCLUSTER DISSOLUTION AND ENERGIES OF METAL IONS DELAY FROM THE SURFACE

Our former studies [25] showed, that atomic oxygen, emerged at the stage of molecular anion O^{2-} decomposition during catalytic electrochemical reduction of molecular oxygen at cathodes of fuel cells, tend to strong chemical bond formation with platinum surface and follows to quasi-chemical oxide because of electron charge transfer from Pt to O. The complex of Pt cation with environment components, partly stabilized, tends to departure off the surface into the solution, with the destruction of the structure itself. The catalytic activity of the binary nanocluster should be decreased. The same situation is observed in the case of carbon compounds oxidation, for instance methanol and ethanol. The surface poisoning occurs under the action of additional components in the environment, especially with the content of chlorine, sulfur and phosphorus. However, the largest problems induce atomic oxygen as extremely active agent in interaction with transition metal nanoclusters.

When curves of activation delay of platinum atoms from the surface were obtained, the following results were collected in the Table 4.

TABLE 4 Activation energies E_A (in kJ/mole) of Pt atom delay from the surface of binary nanoclusters $Pt_{42}Me_{13}$ under the influence of water molecules, hydroxonium ions and dependences on incorporated charges.

Formula of nano-cluster	Without environment	H_2O	H_2O^+ H_3O^+	+1	+2	+3	+4	+5
$Pt_{42}Cr_{13}$	213.015	16.266	15.336	14.402	12.551	11.628	10.445	9.625
$Pt_{42}Fe_{13}$	265.213	18.315	17.434	16.516	15.604	13.632	12.516	11.603
$Pt_{42}Co_{13}$	320.442	30.316	23.512	21.564	19.712	17.522	10.721	14.806
$Pt_{42}Ni_{13}$	286.046	19.522	17.610	8.325	15.829	14.514	10.712	10.828
$Pt_{42}Ru_{13}$	438.383	35.510	34.727	31.486	29.779	27.622	25.815	23.654

As one may conclude from the results, water molecules activate dissolution of platinum atoms. The incorporation of hydroxonium ions as well as positive charges in the system has no influence on the activation energy of platinum decomposition.

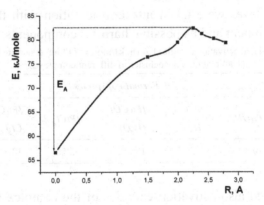

FIGURE 4 Calculation of change of energy interaction $[Pt(OH)\ (H_2O)_4]^+$ with binary nano-clusters $Pt_{42}Co_{13}$ depending on the distance from the surface.

It was shown previously, that oxygen presence is a main factor of possible structural and energy changes, which beside the direct electrochemical reaction may induce by-processes with the lost of catalytic activity of the materials of fuel cell electrodes. After the consideration of adsorption of main molecules and ions from the environment, let's estimate possibilities platinum ion delay from the surface of our nanoclusters. To simplify the consideration we include exclusively water molecules (environment – solution) and hydroxonium ions (active acid environment in the destruction of nanocluster structure). Besides, we try to simulate the cathodic polarization by the known procedure of simple exclusion of addition number of electrons. Barrier activation estimation was realized by reaction way calculations for platinum ion interaction (in the form $[Pt(OH)\ (H_2O)_4]^+$) with binary nanoclusters (Fig. 4).

We considered energies of simple departure of Pt from the surface as cation by means of semiempirical method $PM6$. Activation energies are included into Table 5 in several possible species of complexes. These surface complexes are formed and stabilized by Pt cation interaction with molecules and ions from the environment. Our calculation confirmed existence of prevailing compound $[Pt(OH)\ (H_2O)_4]^+$ (under the condition of exclusion of other harmful ions), while it is accompanied by the largest changes of *Gibbse* isobar-isotherm potential, whereas the activation energy of its departure is the lowest.

Other complexes were taken into consideration with the account of working environment and its possible harmful components.

TABLE 5   Calculated activation energies EA (in kJ/mole) of Pt ion departure from the surface (100) platinum nanocluster in different complexes.

			Formula of species				
	Pt^{+2}	$Pt(OH)^+$	$[Pt(OH)$ $(H_2O)_4]^+$	$[Pt(CO)$ $(H_2O)_4]^{+2}$	$[PtCl_4]^{-2}$	$[Pt(CO)$ $Cl_4]$	$[Pt(CO)$ $Cl_2(H_2O)]$
E_A	120.54	30.826	26.315	18.282	6.256	12.315	14.260

We calculated also activation energies of the complex $[H_2O(PtCl)]^{n+}$ departure from the surface of platinum binary nanoclusters (Figs. 5–6). The obtained results indicate the tendency of their surface destruction by the action of chlorine ions.

In summary the binary nanoclusters $Pt_{42}Cr_{13}$, $Pt_{42}Fe_{13}$, $Pt_{42}Ni_{13}$, $Pt_{42}Ru_{13}$ should be unstable in the environment with the content of hydrated chlorine ions. Their activation energies for surface platinum atom departure are in two times lower, than for pure *Pt* nanocluster. An interesting result was obtained for $Pt_{42}Co_{13}$. Its activation energy is significantly increased.

Thus, in presence of chlorine ions, activation energy of platinum dissolution is sharply lowered. The same statement can be related to the case of carbon compound oxidation, for example methanol, where *CO* forms strong complexes with *Pt* ions with the low activation energy of desorption.

The form of cation departure influence on estimation of environment action on thermodynamics and kinetics of surface processes in oxidation – reduction reactions. So conclusions may be derived about the conditions of fuel cell exploitation and their efficiency.

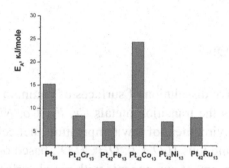

FIGURE 5   Activation energy of $[H_2O(PtCl)]^{n+}$ complexes departure from the surface of *Pt* binary nanocluster.

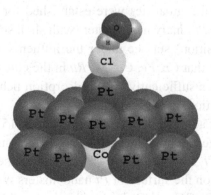

FIGURE 6   Common view of $[H_2O(PtCl)]^{n+}$ complex on the surface of *Pt* binary nanocluster.

Initial data were obtained for comparison purposes in a model vacuum; further calculation showed that inclusion of several water molecules to the system causes decrease in activation energies as much as ten times. The form of departure was picked up as stable $[Pt(OH)\ (H_2O)_4]^+$. As we notice, the most stable nanocluster contains ruthenium, followed by nanocluster with cobalt. Consideration of acid environment leads to partial reduction of activation barriers in all cases, while the exclusion of additional number of electrons from the system (simulation of cathode polarization), causes efficient decrease in activation energies. Such decrease indicates indirectly that fuel cell efficiency with the aim of maximal energy convergence must be regulated with account of accelerated decrease of catalytic activity of nanoclusters in materials of electrodes through their intensive catalytic corrosive destruction.

8.4 CONCLUSIONS

A model of corrosive dissolution of surfaces of platinum binary nanoclusters Pt_nMe_m (Me is the transition metals *Cr, Fe, Co, Ni, Ru*) with shell structure in acid environment of low temperature fuel cells with the content of molecules and ions H_2O, Cl^-, OH^-, H_3O^+, based on calculations of adsorption of environment components with metal surface and activation barriers for dissolution of *Pt* atoms by means of quantum-chemical methods *PM6* and *DFT*.

Physico-chemical regularities were established for structural and energy degradation of *Pt* binary nanocluster (with shell structure of $Pt_{42}Me_{13}$ in different composition) surface under the influence of H_2O, Cl^-, OH^-, H_3O^+. It was shown that *Cr, Fe, Co, Ni, Ru* in the core of $Pt_{42}Me_{13}$ binary nanocluster influence sufficiently on its adsorption behavior and stability to corrosive dissolution, in particular:

– the increase of adsorption heats were depicted (15 – 150%) for water molecules and oxygen atom in the presence of H_3O^+ ions in the series $< Pt_{42}Ru_{13} < Pt_{42}Cr_{13} < Pt_{42}Fe_{13} < Pt_{42}Co_{13} < Pt_{42}Ni_{13}$, with the possible blocking of catalytic centers on the surface of *Pt* nanoclusters with shell structures.

– formation of $Pt(OH)\ (H_2O)_4]^+$ complex is possible on the surface of *Pt* binary nanocluster, where the activation energy of its departure from the surface has the largest values in nanoclusters with Ru and Co in the sequence $Pt_{55} < Pt_{42}Co_{13} < Pt_{42}Ru_{13}$.

– local chemisorption of chlorine ions on the hydrated surface of Pt binary nanoclusters may block catalytic centers of their surface due to high reaction ability of these ions either to dissolution of Pt surface atoms with the formation of $H_2O(PtCl)]^{n+}$ complexes, which possess the lowest activation energy of departure from binary nanocluster surface as compared with pure nanocluster Pt_{55} (more than in twice) with the exclusion of $Pt_{42}Co_{13}$, where an opposite effect was determined (activation energy increase till 60%).

Thus, the surfaces of nanoclusters $Pt_{42}Co_{13}$ and $Pt_{42}Ru_{13}$ with shell structures tend to less degradation in acid environment with the content of water molecules, oxygen atoms and hydroxonium ions, whereas nanocluster $Pt_{42}Co_{13}$ (in addition to similar degradation) is stable to the influence of hydrated chlorine ions in comparison with pure platinum nanocluster.

KEYWORDS

- **core-shell structure**
- **corrosive dissolution**
- **nanoclusters**
- **quantum-chemical modeling**

REFERENCES

1. Zhenmeng Peng, Hong Yang Designer platinum nanoparticles: control of shape, composi-tion in alloy, nanostructure and electrocatalytic property, *"Nano Today"*, **4**, 143–164 (2009).
2. Chien-Te Hsieh, Jia-Yi Lin Fabrication of bimetallic *Pt-M (M = Fe, Co* and *Ni)* nanoparticle/carbon nanotube electrocatalysts for direct methanol fuel cells, *"Journal of Power Sources"*, **188**, 347–352 (2009).
3. E. Antolini, T. Lopes, E. R. Gonzalez An overview of platinum-based catalysts as methanol-resistant oxygen reduction materials for direct methanol fuel cells, *"Journal of Alloys and Compounds"*, **461**, 253–262 (2008).

4. Greeley J., Mavrikakis M. Alloy catalysts designed from first principles, *"Nature Materials"*, **3**, 810–815 (2004).
5. Goddard W., Merinov A., Van Duin A. et al. Multi-paradigm multiscale simulations for fuel cell catalysts and membranes, *"Molecular Simulation"*, **32** (3–4), 251–268 (2006).
6. Shengsheng Zhanga, Xiao-Zi Yuana, Jason Ng Cheng Hina et al. A review of platinum-based catalyst layer degradation in proton exchange membrane fuel cells, *"Journal of Power Sources"*, **194**, 588–600 (2009).
7. Colon-Mercado Hector R., Popov Branko N. Stability of platinum based alloy cathode catalysts in *PEM* fuel cells, *"Journal of Power Sources"*, **155**, 253–263 (2006).
8. Xingwen Yu, Siyu Ye. Recent advances in activity and durability enhancement of Pt/C catalytic cathode in *PEMFC*. Part II: Degradation mechanism and durability enhancement of carbon supported platinum catalyst, *"Journal of Power Sources"*, **172**, 145–154 (2007).
9. A. A. Franco, M. Gerard. Multiscale Model of Carbon Corrosion in a *PEFC*: Coupling with Electrocatalysis and Impact on Performance Degradation, *"J. Electrochem. Soc."*, **155**, B367–B384 (2008).
10. M. Oszcipok, D. Riemann, U. Kronenwett et al. Statistic analysis of operational influences on the cold start behavior of *PEM* fuel cells, *"J. Power Sources"*, **145**, 407–415 (2005).
11. Mukerjee S., Srinivasan S. O_2 reduction and structurerelated parameters for supported catalysts, *"Handbook of Fuel Cells: Fundamental and Applications" / Edited by W. Vielstich, A. Lamm, H. A. Gasteiger. Wiley and Sons.*, **2** (34), 502 (2003).
12. Qing-Song Chen, Shi-Gang Sun, Zhi-You Zhou, Yan-Xin Chen and Shi-Bin Deng CoPt nanoparticles and their catalytic properties in electrooxidation of CO and CH_3OH studied by in situ FTIRS, *"Phys. Chem. Chem. Phys."*, **10**, 3645–3654 (2008).
13. Stamenkovic V. R., Mun B. S., Mayrhoter K. J. J., Ross P. N., Marcovic N. M. Effect of surface composition on electronic structure, stability and electrocatalytic properties of Pt-transition metal alloys: Pt-skin versus Pt-skeleton surfaces, *"J. Am. Chem. Soc."*, **128**, 8813–8825 (2006).
14. Ross P. N. The science of electrocatalysis on bimetallic surfaces, In: *"Elecrocatalysis" / Ed. J. Lipkowski and P. N. Ross, New York, Wiley-VCH*, 43–74 (1998).
15. Demirci U. M. Theoretical means for searching bimetallic alloys as anode electrocatalysts for direct liquidfeed fuel cells, *"J. Power Sources"*, **173**, 11–18 (2007).
16. Tarasevych M. P., Bohdanovskaya V. A. Mechanism of corrosion of the nano-sized multicomponent cathodic catalysts and formation of the core-shell structures, *"Alternative energy and ecology"*, **12** (80), 67–77 (2009) *(in Russian)*.
17. P. Yu, M. Pemberton, P. Plasse. PtCo/C cathode catalyst for improved durability in PEMFCs, "J. Power Sources", **144**, 11–20 (2005).
18. S. Ye, M. Hall, H. Cao, P. He Degradation Resistant Cathodes in Polymer Electrolyte Membrane Fuel Cells, *"ECS Trans"*, **3**, 657–666 (2006).
19. A. A. Turovskiy, V. I. Kopylets, V. I. Pokhmurskiy, S. A. Korniy, G. A. Zaikov Simulation of organic compound oxidation on the surfaces of iron and chromium nanoclusters, In Book: *"Kinetics and Thermodynamics for Chemistry and Biochemistry"*, *Nova Science Publishers, New York*, **2**, 117–124 (2009).

20. V. I. Pokhmursky, S. A. Korniy, V. I. Kopylets' Modelling of the interaction process of oxygen with the surface of binary nanoclusters Pt-Co applying the DFT method, *"Nanosystems, nanomaterials, nanotechnologies"*, **9** (4) (2011) *(in Ukrainian)*.

21. Wang Y., Balbuena P. B. Design of oxygen reduction bimetallic catalysts: Ab-initio derived thermodynamic guidelines, *"J. Phys. Chem."*, **109** B, 18902–18906 (2005).

22. Yasuyuki Ishikawa, Juan J. Mateo, Donald A. Tryk, Carlos R. Cabrera Direct molecular dynamics and density-functional theoretical study of the electrochemical hydrogen oxidation reaction and underpotential deposition of H on Pt(111), *"Journal of Electroanalytical Chemistry"*, **607**, 37–46 (2007).

23. Intermolecular interactions: from two-atomic molecules to biopolymers, *Editor: B. Pyulmann Б. / Moscow.: "Myr"*, 1981, 592 p. *(in Russian)*.

24. S. A. Korniy, V. I. Pokhmurskiy, V. I. Kopylets Quantum chemical and molecular dynamical studies of oxygen interaction with Pt_nX_m and Pd_nX_m (X-first row transition metals) binary nanoclusters, *Proc. 3rd International Symposium "Methods and Applications of Computational Chemistry"*, Odessa (Ukraine), 2009, p. 87.

25. V. Pokhmurskii, S. Korniy, V. Kopylets Computer Simulation of Binary Platinum-Cobalt Nanoclusters Interaction with Oxygen, *"Journal of Cluster Science"*, **22** (3), 449–458 (2011).

26. Lamas E. J., Balbuena P. B. Oxygen reduction on Pd0.75Co0.25 and Pt0.75Co0.25 surfaces: An ab initio comparative study, *"J. of Chem. Theor. and Comp."*, **2**, 1388–1395 (2006).

27. Jing-Shan Do, Ya-Ting Chen, Mei-Hua Lee Effect of thermal annealing on the properties of $Co_{rich core}Pt_{rich shell}$/C oxygen reduction electrocatalyst, *"J. of Power Sourc."*, **172**, 623–632 (2007).

28. Ferrando R., Fortunelli A., Rossi G. Quantum effects on the structure of pure and binary metal nanoclusters, *"Phys. Rev."*, **72** B (8) (2005).

29. Apra E., Fortunelli A. Density-functional calculations on platinum nanoclusters: Pt_{13}, Pt_{38} and Pt_{55}, *"J. Phys. Chem."*, **107** A (16), 2934–2942 (2003).

30. *http:, www.hyper.com*

31. Francesca Baletto and Riccardo Ferrando Structural properties of nanoclusters: energetic, thermodynamic, and kinetic effects, *"Reviews of modern physics"*, 77, 371–423 (2005).

32. James J. P. Stewart Optimization of parameters for semiempirical methods V: modification of NDDO approximations and application to 70 elements, *"J. Mol. Model."*, **13**, 1173–1213 (2007).

33. Stewart J. J. P. Mopac: a semiempirical molecular orbital program, *"J. Comput.-Aided Mol. Des."*, **4** (1), 1–105 (1990).

34. James J. P. Stewart Application of the PM6 method to modeling the solid state, *"J. Mol. Model."*, **14**, 499–535 (2008).

35. Hermann K., Pettersson L. G. M., Casida M. E. et al., *StoBe2009, Version* 2.4 (2009).

36. Becke A. D. Density-functional thermochemistry. III. The role of exact exchange, *"J. Chem. Phys."*, **98** (7), 5648–5563 (1993).

37. J. P. Perdew Accurate Density Functional for the Energy: Real-Space Cutoff of the Gradient Expansion for the Exchange Hole, *"Phys. Rev. Lett."*, **55**, 1665–1668 (1985).

38. A. D. Becke Density-functional exchange-energy approximation with correct asymptotic behavior, "Phys. Rev.", **38** A, 3098–3100 (1988).

39. Lee C., Yang W., Parr R. G. Development of the Colle-Salvetti correlation-energy formula into a functional of the electron density, *"Phys. Rev."*, **37** B, 785–789 (1988).

40. Frumkin A. N. Electrode processes: selected papers, *Moscow: "Nauka"*, 1987, 334 p. *(in Russian)*

41. Kryshtalyk L. I. Electrode reactions. Mechanism of the elementary act, *Moscow: "Nauka"*, 1979, 224 p. *(in Russian)*

42. Atanasyants A. G. Anode behavior of metals, *Moscow: "Metalurgiya"*, 1989, 151 p. *(in Russian)*

CHAPTER 9

EFFECT OF AMINO NAPHTHALENE SULFONIC ACID NATURE ON THE STRUCTURE AND PHYSICAL PROPERTIES OF THEIR COPOLYMERS WITH ANILINE

O. I. AKSIMENTYEVA and V. P. DYAKONOV

CONTENTS

SUMMARY

Paper studied the physical properties and structure of the copolymers based on polyaniline (PANI) and amino naphthalene sulfonic acids (ANSA) with different mutual position of substituents in naphthalene ring, namely, 1-amino-naphthalene-4-sulfonic acid (1,4-ANSA), 1-amino-naphthalene-8-sulfonic acid (1,8-ANSA), 1-amino-naphthalene-5-sulfonic acid (1,5-ANSA) and 1-amino-2-naphthol-4-sulfonic acid. It shown that structure of isomeric ANSA has a significant effect on the thermal stability, magnetic susceptibility and conductivity of the PANI-ANSA systems. The field dependence of magnetization confirms that obtained copolymers are the typical paramagnetic materials. In the interval of $T = 4.2–300$ K all samples demonstrate a stable EPR signal with g-value of 2.0025–2.0034 which is typically for aromatic cation-radicals (polarons). A highest concentration of paramagnetic centers is observed for the (PANI-1,4-ANSA) copolymer (1 : 1) when interaction between ANSA and PANI occurs in ortho-position. In the case of 1,8-ANSA the para-coupling is prevails which causes a higher conductivity and thermal stability of copolymer.

9.1   INTRODUCTION

One way to extend the functionality of the organic materials with electronic conductivity is a chemical or electrochemical copolymerization of different classes of compounds, such as naphthalene derivatives, aromatic and heterocyclic monomers [1–3]. Thus obtained composite materials in the form of thin films or powders have a high electrochemical activity, magnetization, and nonlinear optical properties and can be used in molecular electronics, new sensors and chemical power sources [3–5]. Synthesis the nanocomposites based on the conducting polymer and organic supramolecules such as heteropolyacids or aminonaphthalene sulphonic acids *(ANSA)* acting both as molecular dopants and surfactants, attracts a rising interest because the possibility to formation the nanotubes or nanofibers and polymer nanodispersions during polymerization [3, 5]. For the copolymer of polyaniline *(PANI)* and 2,5-amino naphthalene sulfonic acid a ferromagnetic behavior below $T = 4$ K is observed [6]. However the results on the connection between structure and properties of copolymers of aniline *(An)* with other type of *ANSA* were received not enough.

The process of oxidative polymerization of aromatic amines, including aniline and its derivatives, can occur under the influence of chemical oxidants (persulfate, permanganate, hydrogen peroxide), and anodically polarized electrode during electrochemical synthesis [7, 8]. The cation – radicals forming in result of the splitting of electron from the nitrogen atom subsequently undergo quinoid isomerization making it possible to connect with newly formed radical cations by a type "head-to-tail". Repetition of cycles of oxidation – condensation (coupling) results in a formation of conjugated polymer with electronic conductivity [7]. There is an assumption that cation-radicals of aniline are capable of initiating polymerization of nitro aniline [9] and some amino sulfonic acids, for example, meta-amino benzo sulfonic acid [10], but not participating in the formation of the final product. In the study of the electrochemical behavior of isomeric *ANSA* in aqueous solutions of sulfuric acid was found that these compounds may also undergo a process of oxidative coupling with the participation of amino groups by forming fairly stable radical cations, according to the scheme proposed for the aromatic amines [11]. Given the fact that the potentials of the first wave of anodic oxidation of *ANSA* and aniline ($E = 0.86–0.96$ V, $Ag/AgCl$) are very close to each other [7, 11], we assume the possibility of formation of the coupling products in the form of copolymers during co-oxidation of these compounds. In this paper, we study the relationship between molecular and crystal structure and some physico-chemical properties of aniline copolymers with *ANSA* of different nature, obtained in the conditions of oxidative polymerization.

9.2 EXPERIMENTAL

For synthesis were used the *ANSA* with different mutual position of substituents in naphthalene ring, *namely*, 1-amino-naphthalene-4-sulfoacid (1,4-ANSA), 1-amino-naphthalene-8-sulfoacid (1,8-*ANSA*), 1-amino-naphthalene-5-sulfoasid (1,5-*ANSA*), 1-amino-2-naphthol-4-sulfoacid (*EHT*-acid) and aniline *(An)* in the form of sulfate salt. All reagents were high grade of purity, obtained from *Aldrich*, *ANSA* further were purified by recrystallization [9], ammonium persulfate and sulfuric acid have a highest purity brand. Chemical structure of *ANSA* used for synthesis is presented on *Scheme* 1.

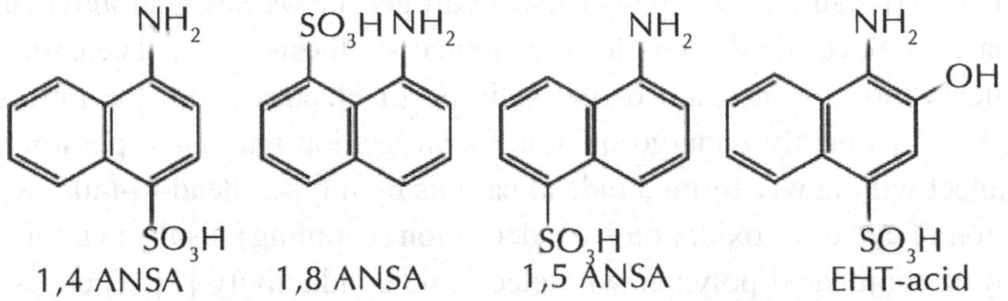

SCHEME 1   Chemical structure of amino naphthalene sulfonic acids

Synthesis and purification of the copolymers was carried out according to Ref. [*12*] in a molar ratio of *ANSA* and aniline 1 : 3 and 1 : 1 in 0.5 *M* solution of sulfuric acid in the presence of ammonium persulfate (molar ratio of amine and peroxide groups was 1 : 1.1).

Molecular structure of composites was investigated by *IR* spectroscopy (*KBr* pellets, *"Specord M-80"*). For study a crystalline structure the method of *X*-ray diffraction was used by means of diffractometer *HZG-4a*, *CoKα*-radiation, $\lambda = 1.7902$ Å, in a discrete mode with step 0.05° 2Θ scan and scanning time 10 s at each point. The calculation of the structure and profile analysis of diffraction patterns were performed using a set of programs *CSD* [*13*]. Thermogravimetric measurements were carried out on the derivatograph *Q-1500 D* in the temperature interval 293–793 K with heating rate of 5 K/min using as a standard Al_2O_3.

Magnetic properties were measured on the *SQEID* magnetometer at T = 4.2 – 293 K as described in [*14*].

Electrical resistance measurements were performed on samples pressed into pellets in a two contacts mode, using a pulsed digital ohmmeter with a measuring range of $1–10^9$ *Ohms*.

EPR spectra of powdered samples of copolymers placed in quartz ampoules, were obtained on the *X*-band spectrometer at $T = 293$ K, used as a reference *DPPH*.

9.3 RESULTS AND DISCUSSION

Oxidative copolymerization of aniline with *ANSAs* was carried out under the influence of a chemical oxidant – ammonium persulfate at room temperature. We found that in the absence of aniline a rate of oxidative polymerization of *ANSAs* is very small, the yield of polymerization product for 72 h just over 18%, while the polymerization of aniline in these conditions leads to almost 84–86% conversion. When coupled with the oxidation of aniline and *ANSAs* yield of the product reaches 67–76% [*12*]. The products obtained in the form of fine-grained black powder were isolated, washed repeatedly with water and acetone and dried under dynamic vacuum for 12 h and analyzed with the a number of physical and chemical methods. According to the elemental and thermogravimetric analysis, the products were a combination of self-doped by sulfuric acid (i.e., doped during synthesis) polyaminoarenes containing chemically bounded sulfonic group. The degree of doping by sulfuric acid in all cases is 30–35 mol. %. We carried out the elemental analysis of de-doped samples (repeatedly washed with a solution of ammonia and water until removed the sulfate groups) for products of copolymerization of *ANSAs* and aniline on the content of sulfonic acid groups. It is revealed that the composition of copolymer is almost identical to that of the original reaction mixture, and only slightly (5%) enriched by polyaniline moieties.

Structure characterization of *PANI-ANSA* systems confirms that all composites consist of the polymerized *ANSA* and polyaniline. According to *IR* – spectroscopy of initial *ANSA*, their polymers and copolymers with aniline was found that in the oxidative polymerization of *ANSA* a main reaction center involved in the initiation and growth of the chain, is the amino group. In the spectra of the monomer *ANSA* are observed absorption bands at 3380 and 3520 cm^{-1}, corresponding to the *NH* in primary aromatic amines [*15*]. As can be seen from the fragments of the *IR* spectra (Fig. 1), for the products of polymerization and copolymerization of *ANSA* with aniline these bands are transformed into a single broad band at 3400–3500 cm^{-1}, characteristic of the secondary amino and imino groups. At the same time, the absorption band responsible for the α-substituted naphthalene ring (3015 cm^{-1}) remains unchanged. In the range of 2000–400 cm^{-1} the absorption bands for deformation vibrations in 1800–1200 cm^{-1} is smoothed, which is typical for the formation of polymers [*15*], and the absorption bands at 1150, 1540–1580 cm^{-1} confirms the formation of secondary amines. For copolymer of 1,8-*ANSA* and aniline was detected the absorption bands which indicates the formation of the products with *para*-coupling (3080, 1150, 830 cm⁻¹).

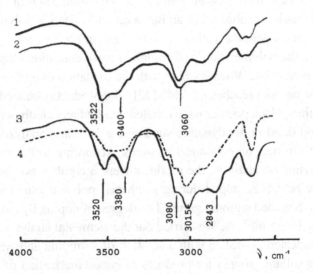

FIGURE 1   Fragments of *IR* spectra for polymer 1,8-*ANSA* (1), monomers 1,8-*ANSA* (2) and 1,4-*ANSA* (3) and poly (1,4-*ANSA* + *An*) copolymer (4).

According to the obtained data one may suggest that the process of oxidative coupling of isomeric *ANSA* and *An* leads to the formation of the copolymer products of different structure. In the case of 1,8-*ANSA* and aniline process may include formation of primary radical cations *ANSA* and aniline, their pair-coupling, the subsequent oxidation of the resulting product in the radical-cation which is capable to further prolongation of the chain:

For copolymers of 1,4- and 1,5-*ANSA* with aniline were recorded the absorption bands at 1150, 758, 778 cm^{-1} (*ortho*-substituted naphthalene cycle). In the region $\upsilon = 1608–1620$ cm^{-1} observed bands corresponding to the deformation vibrations of the *C–C* bond, this bond causes also a presence of the bands with a maximum of absorption at 1450–1456 cm^{-1}. On the basis of the electronic theory of organic reactions [16] and *IR* spectroscopic data, we can assume that the process of oxidative coupling occurs in the *ortho* position, including the formation of *C–C* bonds:

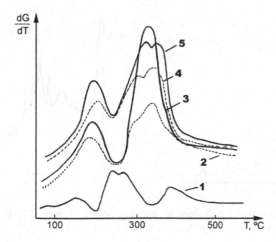

Thermogravimetric measurements showed that each of the resulting copolymer product exhibits a number of peaks in the *DTA* and *DTG* – curves which are not characteristic of individual polymers *ANSA* and aniline (Fig. 2).

In all cases the four major extreme are observed, whose position on the temperature scale depends on the structure of the original *ANSA* taken for synthesis. Endothermic peak at $T = 110–130°C$ associated with desorption of low molecular weight products, in the case of copolymers shifted to $T = 150–180°C$.

FIGURE 2 *DTG* curves for polymer and copolymers of *ANSA* and aniline (1 : 3): 1 – poly(1,8-*ANSA*), 2 – poly(1,4-*ANSA* + *An*), 3 – *PANI*, 4 – poly(1,5-ANSA + *An*), 5 – poly(1,8-*ANSA* + *An*).

Second exothermic peak at 220 – 240°C has a complex profile and corresponds to a series of chemical reactions, are not associated with significant weight loss (oxidation of the amino groups, the formation of cross linked phenazine structures), at 300–320°C occurs an exothermic decomposition of copolymers. For products of coupling the peaks of weight loss shifted to higher temperatures than *ANSA* polymers. In the case of a copolymer of 1,8-*ANSA* with aniline, *DTG* peak position indicating a higher thermal stability of the copolymer in comparison with *PANI*.

In order to study the crystal structure of the polymers was used the method of X-ray powder diffraction. As can be seen from the general form of diffraction patterns, shown on Fig. 3, they are characterized by the presence of sufficient clear peaks of crystalinity over the entire range of scattering angles particularly pronounced in the case of copolymers of 1,8-*ANSA* and aniline. A presence of reflections in the range of small angles (6–10° 2Θ) is typical for systems with long-range order and may indicate the presence of regions with an ordered crystalline structure. With increasing content of aniline in the initial reaction mixture is observed a decrease the degree of crystallinity of the copolymer and in the diffraction patterns along with the reflexes of a crystalline phase is observed well pronounced amorphous halo (Fig. 3*b*).

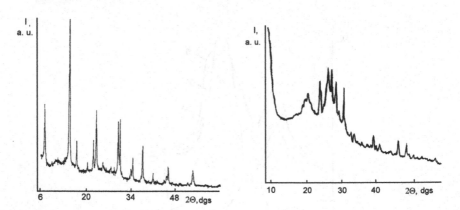

FIGURE 3 X-ray powder diffraction patterns of copolymers 1,8-*ANSA* and aniline, synthesized at a molar ratio of *ANSA* : *An* = 1 : 1 *(a)*; 1 : 3 *(b)*.

Calculation of the parameters of the crystalline phase allowed to assume the existence of an orthorhombic unit cell for a copolymer of 1,8-*ANSA* + *An* (1 : 1), which is confirmed with the data on the structure of the crystalline phase of polyaniline doped with heteropolyacids [17], obtained by means a modified structural model [18]. Calculation of the structure of the copolymer 1.8-*ANSA* with An (1 : 1) allowed us to determine the unit cell parameters: $a = 0.585$; $b = 1.174$; $c = 1.398$ nm. The resulting calculated values of the scattering peaks and the interplanar distances for the most intense reflections showed good agreement with experimental data (Table 1).

TABLE 1 The experimental and calculated parameters of *X*-ray powder diffraction patterns of a copolymer of 1,8-*ANSA* and aniline.

№	$2\Theta_{exp.}$, dgs	$2\Theta_{calcul.}$, dgs	h	k	l	$d_{.exp.}$, Å	$D_{calcul.}$, Å	Δd^{-2}
1	7,274	7,340	0	0	1	14,0873	13,9841	–0,00007
2	14,749	14,710	0	0	2	6,9736	6,9921	+0,00011
3	17,030	17,138	0	1	2	6,0452	6,0075	–0,00034
4	22,150	22,142	0	0	3	4,6597	4,6614	+0,00003
5	32,010	23,010	1	0	2	4,4878	4,4878	+0,00000
6	29,790	29,791	1	1	3	3,4822	3,4821	–0,00001
7	30,390	30,390	0	3	2	3,4150	3,4150	+0,00000
8	37,530	37,530	2	1	1	2,7825	2,7825	+0,00000

Based on the unit cell volume, the data on the density and molecular weight of the proposed elementary unit it was found that the unit cell corresponds to two formula units of the copolymer:

Features of molecular and crystalline structure of the obtained copolymers are also reflected in their macroscopic properties. Influence of the

structure of the isomeric *ANSA* on the physical properties of the copolymers obtained can be traced from the data presented in Table 2. As can be
seen, for all copolymers of *ANSA* and aniline observed elevated values of
conductivity (σ) compared with conductivity of polyaniline obtained in
similar conditions. Such an increase in σ is probably due to the presence in
copolymers structure of chemically bonded sulfonic group with relatively
mobile protons. At the same time the highest value of conductivity is observed for copolymer (1,8-*ANSA* + *An*). The value of σ is 6–8 times higher
than was found for the copolymer products of 1,4- and 1,5-*ANSA* with
aniline and an order of magnitude higher compared to the *PANI*. This copolymer has better thermal stability, characterized by high magnetic susceptibility, not only compared with other copolymers, but with polyaniline
obtained in similar conditions.

TABLE 2   Physical properties of copolymers of aniline with amino naphthalene sulfonic acids (3:1).

Physical Properties	*PANI*	An + 1,8-ANSA	An + EHT	An + 1,4-ANSA	An + 1,5-ANSA
Magnetic susceptibility, $\chi.10^6$, $1/gOe$:					
at $T = 4.2$ K	5	21	19	12	11
at $T = 298$ K	0.25	7.4	6.3	4.8	4.3
Conductivity, 10^{-3}, Cm/cm	0.71	6.4	0.85	1.1	1.4
Mass loss, %:					
at T = 473 K	6.9	6.9	11.3	16.5	13.8
at $T = 493$ K	14.6	14.2	18.4	22.5	19.7
at $T = 573$ K	36.0	22.9	29.0	36.5	30.4

The field dependence of magnetization for *PANI* doped with sulfuric
acid and its copolymers with *ANSA* presented on the Fig. 4 confirms that
obtained copolymers are the typical paramagnetics, and saturated magnetization in applied magnetic field does not achieves.

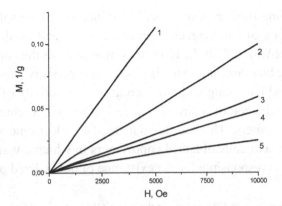

FIGURE 4 Field dependence of magnetization at T = 4.2 K: 1 – poly(1,8-*ANSA+An*), 2 – poly(1,5-*ANSA+An*); 3 – poly(*EHT+An*); 4 – poly(1,4-*ANSA+An*), 5 – *PANI* doped by H_2SO_4.

A magnetic susceptibility of *ANSA-An* copolymers and polyaniline can be seen from Fig. 4. A higher magnetic susceptibility of copolymers of aniline and *ANSAs* consistent with the fact that these compounds at room temperature give a stable in time *EPR* signals. We observed a line of asymmetric shape with the lack of fine structure (Fig. 5), the values of *g*-factor leas in the range of 2.0020–2.0045 which is close to the *g*-factor of aromatic cation-radicals (polarons) [7].

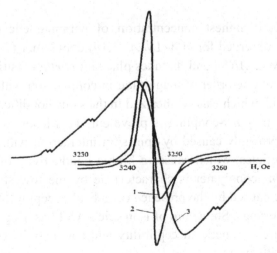

FIGURE 5 *EPR* spectra at T = 298 K of poly(1,8-*ANSA+An*) (1, 3), poly(1,4-*ANSA+An*) (2) copolymers, a molar ratio is 1 : 3 (1) and 1 : 1 (2, 3).

At the same time, a peak-to-peak distance of the signal (ΔHpp) and a concentration of paramagnetic centers *(Ns)* depend on the structure of the original *ANSA* and on the ratio of components in the copolymers (see Table 3). The broadest line shape is observed for copolymers of 1,8-*ANSA* and aniline, which, along with the increased conductivity of these materials may be evidence of a strong delocalization of the charge compared to other copolymers. This fact is confirmed by electronic spectra of the films electrochemically obtained copolymers in which a wide band at λ = 700–750 *nm* is responsible for the existence of delocalized polarons [*19*].

TABLE 3   Parameters of *EPR* spectra of the poly(*ANSA*+*An*) copolymers (T = 293 K).

Copolymer	Molar ratio	g ± 0.0002	ΔHpp, Oe	N_s, g^{-1}
Polyaniline	1 : 0	2.0034	3.1	5.5×10^{19}
(1,4-*ANSA* + An)	1 : 1	2.0028	3.1	1.2×10^{20}
	1 : 3	2.0028	3.1	1.7×10^{19}
(1,8-*ANSA* + An)	1 : 1	2.0026	7.5	3.7×10^{19}
	1 : 3	2.0025	4.0	7.2×10^{19}
(EHT + An)	1 : 3	2.0031	3.2	4.9×10^{18}

However the highest concentration of paramagnetic centers N_s = $1.2 \cdot 10^{20}$ 1/g is observed for (1,4-*ANSA* + *An*) copolymer (1 : 1) when interaction between *ANSA* and *An* takes place into *ortho*-position. The *Ns* is higher by about one order of magnitude in comparison with *PANI* doped by sulfuric acid, which was synthesized in the same conditions. In the case of 1,8-*ANSA* the *para*-coupling is prevalent. This leads to some loss in spin density probably caused by spin-spin interaction with formation of bipolar on particles, which causes a higher conductivity of copolymers. The *(EHT+An)* copolymer is characteristic by the lowest spin density, which may be caused by the presence of radical-acceptor hydroxyl group in naphthalene rings. So, nature of isomeric *ANSA* has a significant effect on the structure, magnetic susceptibility and conductivity of the copolymers *ANSA* with aniline.

9.4 CONCLUSIONS

Comparison of the results on the structure and physico-chemical properties of copolymers of aniline and the *ANSAs* allows us to conclude, the main factor in determining both the structure and functional properties of these materials (conductivity, magnetic susceptibility, thermal stability) is the degree of delocalization of the charge and the length of the conjugation. It reaches its maximum in the case of a copolymer of 1,8-*ANSA* and aniline by allowing the coupling of functional groups in the *para*-position, which leads to the formation of linear chains, which are able to form sufficiently ordered molecular crystals.

At the same time, the process of the oxidative coupling may occurs in the *ortho*-position, as well as the possibility of the formation of simple *C–C* and *N–C* bonds in process of oxidation of 1,4-, 1,5-*ANSA* acid and aniline leads to the formation of oligomer products of different structure, that reflects in their low thermal stability. The high concentration of unpaired spins, characteristic of such products may be due to existence many dangling bonds, which contribute to the spin density at the expense of localized states [20], but do not lead to an increase in conductivity.

ACKNOWLEDGMENTS

This work was partially supported by the Ministry of Science and Higher Education (PL), Grant № 507 492438.

KEYWORDS

- amino naphthalene sulfonic acids,
- copolymers
- magnetic properties
- polyaniline
- structure
- thermal stability

REFERENCES

1. Xiao Y., W.- L. Yu., Pei J., Chen Z., Huang W. and Heeger A.J., *"Synth. Metals."*, **106**, 165–170 (1999).
2. Masdarolomoor F., Innis P. C., Wallace G. G., *"Electrochimica Acta"*, **53**, 4146–415 (2008).
3. Bansal V., Bhandari H., Bansal M. C., Dhawan S. K., *"Indian Journal of Pure and Applied Physics"*, **47**, 667–675 (2009).
4. Saxna S. V., Malhotra B. D., *"Current Applied Physics"*, **3**, 293–305 (2003).
5. Bansal V., Bansal M. C., Dhawan S. K., *"Indian Journal of Engeenering and Material Science"*, **16**, 355–353 (2009).
6. Mizobushi H., Kawai T., Araki H. et al., *"Synth. Metals."*, **69**, 239–240 (1995).
7. Genies E. M., Boyle A., Lapkowski M. and Tsintavis C., *"Synth. Metals."*, **36**, 139–182 (1990).
8. Khastgir B. D., Singhaa N. K., Lee J. H., *"Progress in Polymer Science"*, **34**, 783–810 (2009).
9. Roy B. C., Gupta M. D. and Ray J. K., *"Macromolecules"*, **28** (6) (1995).
10. Roy B. C., Gupta M. D., Browmik L., Ray J. K., *"Synth. Metals."*, **100**, 233–236 (1999).
11. Chernilevskaya G. S., Aksimentyeva E. I., Novikov V. P., *"Theoretical and Applied Chemistry"*, **30** (1), 22–26 (1997).
12. Aksimentyeva O. I., Chernilevskaya G. S., Novikov V. P, Dyakonov V. P., *Patent №* 22046 *(UA), publ.:* 30. 04. 1998, *Bull. №* 2, 4 p.
13. Akselrud L. G., Grin' Yu. N., Zavaliy P. Yu. *et al, Coll. Abstr. of the* XII *European Crystall Meeting*, **3**, p.155, Moscow (1989).
14. Aksimentyeva E. I., Baran M., Dyakonov V. P. et al., *"Physics of Solid State"*, **38** (7), 2277–2285 (1996).
15. Tarutina L. I., Pozniakova F. Yu. Spectral analysis of polymers, *Leningrad: "Khimia"*, p. 33–37 (1986) *(in Russian)*.
16. March J., *"Organic chemistry: reactions, mechanisms and structure"*, **1** (1987), 381 p.
17. Luzny W., Hasik M., *"Solid State Commun."*, **99** (10), 685–689 (1996).
18. Pouget J. P., Jozefowicz M. E., Epstein A. J., Tang X., MacDiarmid A. G., *"Macromolecules"*, **24** (1991), p. 779.
19. Aksimentyeva E. I., *"Electrokhimia"*, **35** (3), 403–406 (1999).
20. Engram D., *"EPR in free radicals"*, Moscow (1961).

CHAPTER 10

UV/VIS-SPECTRA OF SILVER NANOPARTICLES AS CHARACTERISTICS OF THEIR SIZES AND SIZES DISTRIBUTION

A. R. KYTSYA, O. V. RESHETNYAK, L. I. BAZYLYAK, and YU. M. HRYNDA

CONTENTS

SUMMARY

On a basis of the comparative analysis of the references the correlated dependencies between the optical characteristics of aqueous sols of spherical nanoparticles and their diameter have been discovered. As a result, the empirical dependencies between the values of the square of wave frequency in the adsorption maximum of the surface Plasmon resonance and average diameter of the nanoparticles were determined as well as between the values of the adsorption band width on a half of its height and silver nanoparticles distribution per size. Proposed dependencies are described by the linear equations with the correlation coefficients 0.97 and 0.84, respectively.

10.1 INTRODUCTION

An exponential growth in a field of the fundamental and applied sciences connected with a synthesis of the nanoparticles of noble metals, studies of their properties and practical application is observed for the last decades. Such rapid development of the scientific investigations in the above said fields is caused first of all by the development of the instrumental and synthetic methods of obtaining and investigations of such materials in connection with their wide using in microelectronics, optics, catalysis, medicine, sensory analysis and etc. Silver nanoparticles *(Ag-NPs)* are characterized by unique combination of the important physical-chemical properties, *namely* by excellent optical characteristics, by ability to amplify the signal in spectroscopy of the combination dispersion [1], and also by high antibacterial properties. Among the all metals possessing by the characteristic phenomenon of the surface Plasmon resonance *(SPR)* exactly the silver is characterized by the greatest efficiency of the Plasmon's excitation that leads to the abnormally high value of the extinction coefficient of *Ag-NPs* [2]. Under conditions of modern tendency to the miniaturization and the necessity to improve the technological processes of the new materials obtaining based on *Ag-NPs*, there is a problem of their identification, which requests the cost equipment and causes a search of the alternative ways of their average size and size distribution determination by others methods, *in particular*, by calculated ones with the use of the empirical equations and dependencies which are based on the property of the adsorption of the electromagnetic irradiation in *UV*/visible diapason by the sols of *Ag-NPs* [3].

10.2 THEORETICAL GROUNDS

Generally, for theoretical description of the *SPR* phenomenon of the metallic little particles and for *Ag-NPs*, in particular, the solving's of the *Maxwell's* equation are used, which in 1908 have been proposed by *Gustav Mi* [4]. The extinction coefficient (C_{ext}) of the spherical nanoparticles in accordance with Mi's theory is described by the equation:

$$C_{ext} = \frac{24\pi^2 r \varepsilon_M^{3/2}}{\lambda} \frac{\varepsilon_2}{\left(\varepsilon_1 + 2\varepsilon_M\right)^2 + \varepsilon_2^2} \tag{1}$$

where r is the radius of a particle, l is a length of a wave of the electromagnetic irradiation, ε_M is the dielectric transmissivity of the solvent, ε_1 is a real part of the value of dielectric transmissivity of a part of the metal, ε_2 is the imagined part of the value of dielectric transmissivity of a part of the metal.

It is known [3, 5] that the position of *SPR* maximum adsorption depends on a size of the *Ag-NPs*. Such phenomenon is explained by dependence of real and imagined parts of the dielectric permeability of silver on size of the nanoparticle. In accordance with *Drude's* model [6], ε_1 and ε_2 can be described by the expressions:

$$\varepsilon_1 = \varepsilon'_{bulk} + \frac{\omega_p^2}{\omega^2 + \omega_d^2} - \frac{\omega_p^2}{\omega^2 + \omega_r^2}, \tag{2}$$

$$\varepsilon_2 = \varepsilon''_{bulk} + \frac{i\omega_p^2 \omega_r}{\omega\left(\omega^2 + \omega_r^2\right)} - \frac{i\omega_p^2 \omega_d}{\omega\left(\omega^2 + \omega_d^2\right)}, \tag{3}$$

$$\omega_r = \omega_d + \frac{v_F}{r}, \tag{4}$$

where ε'_{bulk} and ε''_{bulk} are values of the real and of the imagined parts of dielectric permeability of silver mass, ω, ω_p and ω_d are correspondingly the frequency of the electromagnetic irradiation, Plasmon frequency of the metal and decrement of electron gas extinction in the mass metal, v_F is the *Fermi* rate.

However, calculated accordingly to such expressions adsorption spectra of aqueous sols of spherical *Ag-NPs* are differed from the experimental ones, that can be explained by different reasons, *in particular*: firstly, in presented example of the calculations it was not taken into account the distribution of *Ag-NPs* per sizes, that has an influence on a value of the *SPR* adsorption band width on a half of its height and, secondly, in classical *Drude's* model the adsorbed stabilizer on the surface doesn't take into account; in turn, such stabilizer can influence on the value of the wave length in adsorption maximum of the *Ag-NPs* sol.

In order to determine the dependencies between the optical characteristics and size of the nanoparticles we have done an analysis of the great data of references [5, 7–42] concerning to the synthesis and the investigations of *Ag-NPs*.

10.3   RESULTS AND DISCUSSION

It was determined (see Fig. 1), that a square of the wave frequency in adsorption maximum of *SPR* (ω^2) linearly depends on a value of the average diameter (d) *Ag-NPs*.

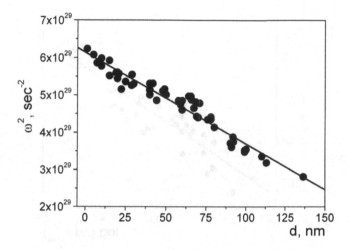

FIGURE 1 Dependence between the square of the wave frequency in adsorption maximum of *SPR* and diameter of *Ag-NPs*.

Such dependence is described by the expression:

$$\omega^2 = (6,14 \pm 0,05) \cdot 10^{29} - (2,45 \pm 0,08) \cdot 10^{27} \qquad (5)$$

with the correlation coefficient 0.97.

At the same time, it was not discovered the direct dependence between the width of the adsorption band of *Ag-NPs* on a half of its height (Δl) and nanoparticles distribution per size (Δd). Evidently, it is connected with the nonmonotonic change of the adsorption band of *Ag-NPs* at their size increasing [6]. However, the all analyzed data are satisfactory described by the linear equation:

$$\log(d \cdot \Delta\lambda) = (0,2 \pm 0,1) + (0,89 \pm 0,06) \cdot \log(\Delta d \cdot \lambda_{max}) \qquad (6)$$

with the correlation coefficient 0.84 (see Fig. 2). Here λ_{max} is a value of the wavelength in a maximum of the *SPR*.

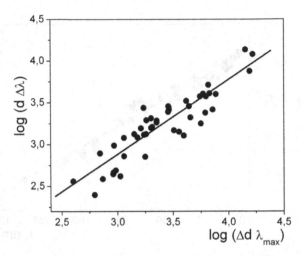

FIGURE 2 Dependence between the logarithms of compositions $\log(d \cdot \Delta\lambda)$ and $\log(\Delta d \cdot \lambda_{max})$ for *Ag-NPs*.

It is necessary to notify, that in processed data *Ag-NPs* were obtained in aqueous solution with the use of different upon nature stabilizers of the surface and precursors. However, in spite of this fact, discovered by us dependencies are good described with the respective correlation coefficients. It is clear, that for the explanation of nature for such dependencies the advanced theoretical analysis of the electron gas interaction with the electromagnetic irradiation is necessary, however, at the presented stage, such empirical dependencies can be used as the rapid method of the synthesized by different methods of *Ag-NPs* identification in laboratory and industrial conditions without the application of complicated, cost and often absent in Ukraine devices for their identification.

10.4 CONCLUSIONS

Empirical dependencies between the dimensional and optical characteristics of silver nanoparticles were determined. Such dependencies can be used for estimation of value of the average diameter and distribution per size of *Ag-NPs* without application of complicated equipment.

KEYWORDS

- extinction spectra
- silver nanoparticles
- surface Plasmon resonance
- UV/VIS-spectra

REFERENCES

1. Yu. A. Krutyakov, A. A. Kudrinsky, A. Yu. Olenin, G. V. Lisichkin Synthesis and Properties of Silver Nanoparticles: Achievements and Perspectives, *"Uspiekhi Khimiji"*. 2008, v. 77. № 3. 242–269 *(in Russian)*.

2. S. Z. Malynych Synthesis and Optical Propertties of Superdispersion ver Aqueous Suspensions, *Journal of Physical Studies*. 2009, v. 13, № 1, 18011–18016.

3. S. Z. Malynych Estimation of Size and Concentration of Silver Nanoparticles in Aqueous Suspensions from Extinction Spectra, *J. Nano- Electron. Phys.* 2010, v. 2, № 4, 5–11.

4. G. Mie Beiträge zur Optik trüber Medien, speziell kolloidaler Metallösungen, *Annalen der Physik*. 1908, v. 330 (3), 377-445.

5. D. D. Evanoff, G. Chumanov Synthesis and Optical Properties of Silver Nanoparticles and Arrays, *Chem. Phys. Chem.* 2005, v. 6, 1221–1231.

6. A. Slistan-Grijalva, R. Herrera-Urbina, J. F. Rivas-Silva, M. Avalos-Borja, F. F. Castillo-Barraza, A. Posada-Amarillas Classical Theoretical Characterization of the Surface Plasmon Absorption Band for Silver Spherical Nanoparticles Suspended in Water and Ethylene Glycol, *Physica, E.* 2005, v. 27, 104–112.

7. A. I. Kryukov, N. N. Zinchuk, A. V. Korzhak, S. Ya. Kuchmiy An Influence of the Conditions of Catalytic Synthesis of Nanoparticles of Metallic Silver on Their Plasmon Resonance, *Theoret. i Experim. Khimiya.* 2003, v. 39, № 1, 8–13 *(in Russian)*.

8. V. V. Bryukhanov, N. S. Tikhomyrova, R. V. Gorlov, V. A. Sliezhkin An Interaction of Surface Plasmons of Silver Nanoparticles on Silokhrom with Electron-Excited Adsorbates of the Molecules of Rhodamine 6ZH, *Izviestiya KGTU.* 2011, № 23, 11–17 *(in Russian)*.

9. S. N. Munro, W. E. Smith, M. Garner, J. Clarkson, P. C. White Characterization of the Surface of a Citrate-Reduced Colloid Optimized for Use as a Substrate for Surface-Enhanced Resonance Raman Scattering, *Langmuir.* 1995, v. 11, 3712–3720.

10. L. N. Podlegayeva, D. M. Russakov, S. A. Sozinov, T. V. Morozova, I. L. Shvayko, N. S. Zvidentsova, L. V. Kolesnikova Investigations of Properties of Silver Nanoparticles Obtained by Reduction From the Solutions and by the Thermal Deposition in Vacuum, *"Vestnik Kiemierov-skiego gossudarstviennogo universitieta"*. 2009, iss. 2(38), 91–96 *(in Russian)*.

11. V. Parashar, R. Parashar, B. Sharma, A. Pandey Parthenium Leaf Extract Mediated Synthesis of Silver Nanoparticles: a Novel Approach Towards Weed Utilization, "*Digest Journal of Nanomaterials and Biostructures*". 2009, v. 4, № 1, 45-50.

12. Y. Yin, Z.–Yu. Li, Z. Zhong, B. Gates, Y. Xia, S. Venkateswaranc Synthesis and Characterization of Stable Aqueous Dispersions of Silver Nanoparticles Through the Tollens Process, "*J. Mater. Chem.*" 2002, v. 12, 522–527.

13. Y. Sun, Y. Xia Gold and Silver Nanoparticles: A Class of Chromophores with Colors Tunable in the Range From 400 to 750 nm, "*Analyst.*" 2003, v. 128, 686–691.

14. H. Wang, X. Qiao, J. Chen, S. Ding Preparation of Silver Nanoparticles by Chemical Re-duction Method, "*Colloids and Surfaces A: Physico-chem. Eng. Aspects*". 2005, v. 256, 111–115.

15. Y. M. Mohan, K. Lee, T. Premkumar, K. E. Geckeler Hydrogel Networks as Nanoreactors: A Novel Approach to Silver Nanoparticles for Antibacterial Applications, "*Polymer*". 2007, v. 48, 158–164.

16. X. Li, J. Zhang, W. Xu, H. Jia, X. Wang, B. Yang, B. Zhao, B. Li, Yu. Ozak Mercaptoacetic Acid-Capped Silver Nanoparticles Colloid: Formation, Morphology, and *SERS* Activity, "*Langmuir*", 2003, v. 19, № 10, 4285–4290.

17. N. Vigneshwaran, R. P. Nachane, R. H. Balasubramanya, P. V. Varadarajan A Novel One-Pot "Green" Synthesis of Stable Silver Nanoparticles Using Soluble Starch, "*Carbohydrate Research*". 2006, v. 341, 2012–2018.

18. S. Shankar, A. Rai, A. Ahmad, M. Sastry Rapid Ыlynthesis of *Au, Ag*, and Bimetallic *Au* core *Ag* Shell Nanoparticles Using Neem (Azadirachta Indica) Leaf Broth, "*Journal of Colloid and Interface Sci.*" 2004, v. 275, 496–502.

19. S. Chandran, M. Chaudhary, R. Pasricha, A. Ahmad, M. Sastry Synthesis of Gold Nanotriangles and Silver Nanoparticles Using Aloe Vera Plant Extract, "*Biotechnol. Prog.*" 2006, v. 22, 577–583.

20. H. Hiramatsu, F. Osterloh A Simple Large-Scale Synthesis of Nearly Monodisperse Gold and Silver Nanoparticles with Adjustable Sizes and With Exchangeable Surfactants, "*Chem. Mater.*" 2004, v. 16, № 13, 2509–2511.

21. J. Y. Song, B. Kim Rapid Biological Synthesis of Silver Nanoparticles Using Plant Leaf Extracts, "*Bioprocess Biosyst Eng.*" 2009, v. 32, 79–84.

22. S. Li, Yu. Shen, A. Xie, X. Yu, L. Qiu, L. Zhang, Q. Zhang Green Synthesis of Silver Nanoparticles Using Capsicum Annuum L. Extract, "*Green Chem.*" 2007, v. 9, 852–858.

23. J. Zhu, S. Liu, O. Palchik, Yu. Koltypin, A. Gedanken Shape-Controlled Synthesis of Silver Nanoparticles by Pulse Sonoelectrochemical Methods, "*Langmuir*". 2000, v. 16, 6396-6399.

24. A. R. Shahverdi, S. Minaeian, H. R. Shahverdi, H. Jamalifar, A. A. Nohi Rapid Synthesis of Silver Nanoparticles Using Culture Supernatants of Enterobacteria: A Novel Biological Approach, "*Process Biochemistry*". 2007, v. 42, 919–923.

25. G. A. Martinez-Castanon, N. Nino-Martinez, F. Martinez-Gutierrez, J. R. Martinez-Mendoza, F. Ruiz Synthesis and Antibacterial Activity of Silver Nanoparticles With Different Sizes, "*J. Nanopart Res.*" 2008, v. 10, 1343–1348.

26. A. Shahverdi, A. Fakhimi, H. Shahverdi, S. Minaian Synthesis and Effect of Silver Nanoparticles on the Antibacterial Activity of Different Antibiotics Against Staphylococcus Aureus and Escherichia Coli, "*Nanomedicine: Nanotechnology, Biology and Medicine*". 2007, v. 3, iss. 2, 168–171.

27. H. Huang, X. Yang Synthesis of Polysaccharide-stabilized Gold and Silver Nanoparticles: A Green Method, *"Carbohydrate Research"*. 2004, v. 339, iss. 15, 2627–2631.
28. A. Ahmad, P. Mukherjee, S. Senapati, D. Mandal, M. I. Khan, R. Kumar, M. Sastry Extracellular Biosynthesis of Silver Nanoparticles Using the Fungus Fusarium Oxysporum, *"Colloids and Surfaces B: Biointerfaces"*. 2003, v. 28, iss. 4, 313–318.
29. U. Nickel, K. Mansyreff, S. Schneider Production of Monodisperse Silver Colloids by Reduction With Hydrazine: the Effect of Chloride and Aggregation on *SER(R)S* Signal Intensity, *"J. Raman Spectrosc."* 2004, v. 35, 101–110.
30. http://www.sigmaaldrich.com/materials-science/nanomaterials/silver-nanoparticles.html
31. H. H. Huang, X. P. Ni, G. L. Loy, C. H. Chew, K. L. Tan, F. C. Loh, J. F. Deng, G. Q. Xu Photochemical Formation of Silver Nanoparticles in Poly(N-vinylpyrrolidone), *Langmuir"*. 1996, v. 12, 909-912.
32. A. Henglein, M. Giersig. Formation of Colloidal Silver Nanoparticles: Capping Action of Citrate, *"J. Phys. Chem. B."* 1999, v. 103, 9533-9539.
33. http://www.nanocomposix.com/products/silver/spheres
34. O. V. Dementyeva, A. V. Malkovsky, M. A. Philippenko, V. M. Rudoy Comparative Investigation of Properties of Silver Hydrosols Obtained by Citrate and Citrate sulfate Methods, *"Colloidal Journal"*. 2008, v. 70, № 5, 607-619 *(in Russian)*.
35. V. Sharma, R. Yngard, Y. Lin Silverna Nanoparticles: Green Synthesis and Their Antimicrobial Activities, *"Adv. in Colloid and Interface Sci."* 2009, v. 145, 83–96.
36. R. M. Tilaki, A. Irajizad, S. M. Mahdavi Stability, Size and Optical Properties of Silver Nanoparticles Prepared by Laser Ablation in Different Carrier Media, *"Appl. Phys. A."* 2006, v. 4, 215–219.
37. A. Panacek, L. Kvitek, R. Prucek, M. Kolar, R. Vecerova, N. Pizurova, V. K. Sharma, T. Nevecna, R. Zboril Silver Colloid Nanoparticles: Synthesis, Characterization, and Their Antibacterial Activity, *"J. Phys. Chem. B."* 2006, v. 110, 16248–16253.
38. D. C. Tien, C. Y. Liao, J. C. Huang, K. H. Tseng, J. K. Lung, T. T. Tsung, W. S. Kao, T. H. Tsai, T. W. Cheng, B. S. Yu, H. M. Lin, L. Stobinski Novel Technique for Preparing a Nano-Silver Water Suspension by the Arc-Discharge Method, *"Rev. Adv. Mater. Sci."* 2008, v. 18, 750–756.
39. J. Liu, R. Hurt Ion Release Kinetics and Particle Persistence in Aqueous Nano-Silver Colloids, *"Environ. Sci. Technol."* 2010, v. 44, 2169–2175.
40. D. L. Van Hyning, C. F. Zukoski Formation Mechanisms and Aggregation Behavior of Borohydride Reduced Silver Particles, *"Langmuir"*. 1998, v. 14, 7034–7046.
41. S. D. Solomon, M. Bahadory, A. V. Jeyarajasingam, S. A. Rutkowsky, C. Boritz Synthesis and Study of Silver Nanoparticles, *"Journal of Chemical Education"*. 2007, v. 84, № 2, 322–325.
42. E. V. Abhalimov, A. A. Parsayev, B. G. Ershov Obtaining of Silver Nanoparticles in Aqueous Solutions in the Presence of Stabilizing Carbonate-Ions, *"Colloidal Journal"*. 2011, м. 73, № 1, 3–8.

CHAPTER 11

SYNTHESIS AND STUDIES OF THE ANTICORROSION ACTIVITY OF THE ZINC PHOSPHATE NANOPARTICLES

L. I. BAZYLYAK, A. R. KYTSYA, I. M. ZIN', and S. A. KORNIY

CONTENTS

SUMMARY

The new methods of the nanosized inhibited pigments synthesis have been developed based on the zinc phosphate and/or (poly)phosphate. This permits to create the new competitive materials for their application in paint and varnish industry under the industrial conditions. It was shown, that the acrylic monomers can be used as the effective modifiers of the zinc phosphate nanoplates surface [1]. At studying of the monomer nature and its concentration influence on a size and on a form of the obtained nanoplates it was determined, that at the butyl methacrylate using as the surface's modifier the plates by the average size 200 ± 70 nm and by the thickness < 20 nm can be obtained. With the use of the electrochemical impedance spectroscopy the anticorrosion activity of the obtained pigment in composition of the undercoatings [2] was investigated. It was determined, that at the addition of 1% mass. of the zinc nanophosphate the efficiency of the undercoating is higher in comparison with the sample containing of 5% mass of the "Novinox PZ02" pigment.

11.1 INTRODUCTION

The metals, which are used by humanity in different fields of the industrial production and technics, and also the different prefabricated metals are always corroded, since there are factors in the environment which accompany of the corrosion processes proceeding, *in particular,* the oxygen of air, humidity, dust, etc.

One among widely distributed methods of the metals and also the different prefabricated metals protection from the corrosion *is an application of the anticorrosive coatings.* A role of the corrosion inhibitors in such coatings plays the pigments, which besides the protective functions give also the color and good covering ability to the coatings. A property of the pigments to brake the corrosive process at the boundary "metal-coating" is caused by the backing-up of the anodic, cathodic or simultaneously of the both processes of the electrochemical corrosion, which is achieved at the expense of the disengagement ions by the pigment, which form the passivation layers on the surface of metals, pH increasing at the boundary "film- substrate" and etc.

Modern demands to the ecological safety of production at the simulta-neous decreasing of the unit value during the obtaining of the protective coating possessing by the anticorrosive properties cause the necessity to search the new ecologically pure and cheaper pigments. The phosphates were the first compounds among those which were used for the decreasing of the toxicity of the anticorrosive coatings instead of the widely applied chromium- and stannum-containing ones. Zinc phosphates and chromi-um phosphates, which represent by themselves the nontoxic crystalline hydrates [3] mainly were used as the phosphates-containing pigments. Zinc phosphate $Zn_3(PO_4)\cdot nH_2O$ is characterized by a weak solubility in water, however it is easy dissolved in the acids. Chromium phosphate $Cr(PO_4)\cdot nH_2O$ is practically insoluble in water, is stable to the action of the acids and alkalies and it is not used as the self-contained anticorrosive pigment. It is used in the pigment compositions, *in particular*, in the chro-mate ones, in which the synergistic result is observed, which is explained by the increase of the chromate solubility and by the intensification of their anticorrosive properties [4]. Besides of the listed above salts of the phosphoric acid, the anticorrosive properties demonstrate also the others salts, including the acidic phosphates of the *Aluminum, Barium, Calcium, Magnesium, Manganese* [5]. The condensed phosphates of metals are also used for the protection of metals from the corrosion [6], namely: diphos-phates of *Copper* $Cu_2P_2O_7$, *Calcium* $Ca_2P_2O_7$, *Magnesium* $Mn_2P_2O_7$; poly-phosphates of *Calcium* $Ca_3(P_3O_{10})_2\cdot1,5H_2O$, *Zinc* $Zn_3(P_3O_{10})_2\times H_2O$, *Alu-minum* $Al_3(P_3O_{10})_2\cdot2H_2O$; cyclotetraphosphates of *Iron* $Fe_2P_4O_{12}$, *Copper* $Cu_2P_4O_{12}$, *Nickel* $Ni_2P_4O_{12}$, *Zinc* $Zn_2P_4O_{12}$, *Magnesium* $Mg_2P_4O_{12}$, *Calcium* $Ca_2P_4O_{12}$, *Manganese* $Mn_2P_4O_{12}$. Condensed phosphates of metals possess by the better anticorrosive properties, since the ions which are formed at their hydrolysis are characterized by the well-defined expressed complex-forming ability relatively to the ion Fe^{3+}, than to the ion PO_4^{3-}. The surface layer of compound, which is formed as a result of the complex-forming reactions, For example, by general formula $Fe_xAl_y(PO_4)_z$, makes the sur-face of steel by inert [7].

However, *a general disadvantage of the phosphate pigments used in the anticorrosive coatings, is a low efficiency of the under-film corrosion process evolution on the initial stages,* which is connected with their low water solubility, and, obtained accordingly to the described above methods zinc phosphates particles represent by themselves the crystals by the micron size. Nanosized particles are characterized by the anomalous physical-chemical properties in comparison with the heavy simple of a compound by the same analogous chemical composition. *In particular*, at the transition to the ultra-dispersive particles there are changes in melting temperature, electronic, magnetic, optical, catalytic and others properties of the compound.

The main advantage of the nanosized anticorrosive pigments, in comparison with the traditionally existing ones by the micron size, is the ratio of a high surface area to the volume, and the anticorrosive properties of the coating based on the nanosized pigments directly proportionally depend on a size of the particles. An application of the nanopigments permits to improve simultaneously the thermal stability, durability at the stroke, chemical stability, barrieric and anticorrosion characteristics. In connection with this fact, the actual is the search of the nanosized pigments synthesis techniques, *in particular*, of the zinc phosphate nanosized crystals. In references there is separate information about the methods of the zinc phosphates nanosized crystals synthesis, which in general, has the episodical character. Evidently, this is connected with the complication of the nanosized inorganic salts in aqueous solutions obtaining, which is explained by the fact that in spite of the very low solubility of zinc phosphate in water, in accordance with the *Kelvin's* equation $\frac{C}{C_0} = \exp\left(\frac{2\sigma V}{rRT}\right)$ the less crystals are dissolved and the big ones are aggregated.

That is why *the aim of the presented work was* to develop the methods of a synthesis of the nanosized pigments with anticorrosive properties on a basis of the zinc phosphates, which permit to create under the industrial conditions the new competitive materials for their using in varnished-paint industry, for protection of the metal construction from the corrosion, etc.

The nanosized pigments with the anticorrosion properties based on the zinc phosphates *were used as the investigation object.*

A synthesis of the nanosized pigments on a basis of zinc phosphates *was the subject of the investigation* with the use of the chemical methods of synthesis, electron microscopy and electrochemical impedance spectroscopy.

11.2   EXPERIMENTAL

11.2.1   STARTING REAGENTS

The following starting reagents were used for the synthesis of nanosized pigments with the anticorrosive properties on a basis of the zinc phosphates, *namely: zinc acetate* $Zn(CH_3COO)_2 \cdot 2H_2O$ *(State standard specification* 4174–77) from the *"Sfera Sim"*™ with the content of the main substance [3] 99.5% and *ammonium phosphate* $(NH_4)PO_4$ *(State standard specification* 4174–78) from the *"Sfera Sim"*™ with the content of the main substance [3] 99.0%.

For the modification of surface, for the stabilization and for the improvement of the compatibility of zinc phosphate nanoparticles with the polymeric carrier (lacquer) the following reactive monomers were used, namely: butylmethacrylate *(BMA)* $(H_2C=C(CH_3)COOC_4H_9$ by specification: *Mn* 142.2; $T_{boil.}$ 436 *K*; d_4^{20} = 0.8950 and the content of the main substance 99.0%) and butylacrylate *(BA)* $(H_2C=CHCOOC_4H_9$ by specification: *Mn* 128.17; $T_{boil.}$ 420.4 *K*; d_4^{20} = 0.8935 and the content of the main substance 99.0%), emulsifier E-30 by formula which represents by itself the mixture of the linear alkansulfonates by general formula $R\text{-}SO_3Na$ (where *R* corresponds to the hydrocarbon chain by average length C_{15}) (*Mn* 318; $T_{melt.}$ 413 *K*; the content of the main substance 93.0%), and also the polyvinyl alcohol and polyethylene glycol as the stabilizers.

Synthesis of the phosphate pigments was carried out in reactive medium water / methanol at their volume ratio 50 : 50 (methanol CH_3OH was used by specification *Mn* 128,17; $T_{boil.}$ 338 *K*; d_4^{20} = 0.7918 and the content of the main substance 99,0%).

11.2.2   METHODS OF THE EXPERIMENTS CARRYING OUT

Synthesis of nano- and ultradispersive particles of zinc phosphate was carried out accordingly to the ion change reaction $3Zn^{2+} + 2PO_4^{3-} = Zn_3(PO_4)_2 \times 4H_2O\downarrow$ in glass reactor under the room conditions in medium water / methanol at their volume ratio 50 : 50. For the reagents stirring the magnetic or mechanical stirrer and / or ultrasonic bath *"Crystall-2"* has been used. The control of the pH medium was carried out with the use of the laboratory ion-meter *"Expert-001"*. Synthesized product was separated from the reactive mixture by filtration with the use of the *Bunsen* flask and *Buchner* bailer and was dried at 120°C.

Form, size and the elementary composition of the synthesized ultradispersive and nanosized zinc phosphates were estimated with the use of the scanning electron microscope EV*O-40XVP* with the system of the microanalysis *INCA Energy* 350.

Dispersive analysis of the anticorrosive pigments was carried out in accordance with the sedimentation method in the gravitational field [8].

Anticorrosion effectiveness of nanophosphate pigments in alkyd coatings on *D*16T aluminum alloy were studied by electrochemical impedance spectroscopy *(EIS)*. The samples were defected intentionally by drilling holes of 1 mm diameter in the coating to expose the underlying substrate and to reveal the effect of inhibitors on its under film corrosion. The measurements were carried out close to corrosion potential using Gill AC potentiostat, a plexiglass tube cell glued to the surface of the sample, a saturated *Ag/AgCl* reference electrode and a platinum auxiliary electrode in the current frequency range from 1 k*Hz* to 0.1 Hz and with the signal amplitude of 20 *mV*. The working area of the samples was 5.0 cm².

11.3   RESULTS AND DISCUSSION

11.3.1   SYNTHESIS OF NANOSIZED ZINC PHOSPHATE IN AQUEOUS MEDIUM

The main problem at the synthesis of the salts metals nanoparticles is fact that in accordance with the *Kelvin's* equation $\frac{C}{C_0} = \exp\left(\frac{2\sigma V}{rRT}\right)$ at the size of the particle of substance decrease its solubility is increased. This means that at the synthesis of the nanoparticles of weakly soluble salts it is necessary to block the transfer of the substance's mass from the less particles to the larger ones. Usually, such problem is solved by the application of different surface-active substances, which are adsorbed on the surface of particles and blocks of it. We have studied an influence of different upon nature surface-active substances (emulsifiers E-30 and TX-100), and also the stabilizers (polyvinyl alcohol and polyethylene glycol) on a form and on a size of obtained crystals of zinc phosphate. It was determined (Fig. 1), that the obtained particles represent by themselves the plates by thickness of 0.1–1 µm and by the length of 3–10 µm.

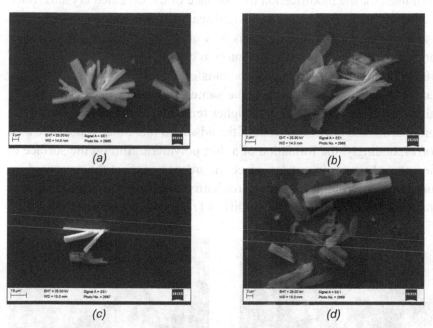

FIGURE 1   *SEM*-images of the zinc phosphate samples obtained in the presence of E-30 *(a)*, TX-100 *(b)*, polyvinyl alcohol *(c)* and polyethylene glycol *(d)*.

Obtained results mean, that such surface-active substances cannot effectively inhibit the process of the mass transfer between the particles. Probably this is connected with a good solubility of emulsifies in water.

11.3.2   SYNTHESIS OF NANOSIZED ZINC PHOSPHATES IN WATER-ORGANIC MEDIUM

One method among the all well-known ones to prevent the solubility of the zinc phosphates nanosized crystals is to carry out the reaction of their synthesis in the mix of the organic solvents with water, that permits to decrease the solubility of zinc phosphate in reactive medium and also effectively to block the mass transfer between the formed particles. That is why reaction was carried out in water-methanol mixture, and acrylic monomers (*namely*, butylacrylate *(BA)* and butylmethacrylate *(BMA)*) have been used for the modification of a surface of the obtained crystals. Such monomers are characterized by insignificant surface-active properties, are weakly soluble in water, and that is why at the new phase formation their molecules will be adsorbed on a surface of zinc phosphate nanoparticles and additionally to inhibit the mass transfer of the substance from the less particles to the greater ones. At the same time, under drying of the obtained product of the reaction at higher temperature the proceeding of the ion-coordinated polymerization of the adsorbed monomer is possible and, correspondingly, the formation of a thin polymeric film on the surface of zinc phosphate particles takes place. In order to confirm the presented assumptions, the samples of zinc phosphate in the presence of *BA* and *BMA* and also in the absence of the modifiers [1] have been synthesized.

(a)

(b)

FIGURE 2   *(Continued)*

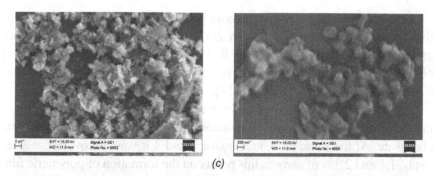

(c)

FIGURE 2 *SEM* images of zinc phosphate nanoparticles obtained in water-methanol mixture without modifying agent of the surface *(a, sample № 1)*, in the presence of *BA (b, sample № 2)* and in the presence of *BMA (c, sample № 3)*.

As we can see from the presented Fig. 2, obtained nanoparticles represent by themselves the plates by thickness > 100 *nm* under their synthesis in absence of the modifying agent, by thickness < 50 nm in a case of their synthesis with the use of *BA* and by thickness < 20 nm in a case of their synthesis in the presence of *BMA*.

On a basis of the presented data the average linear sizes of the obtained zinc phosphate nanoplates were calculated (see Fig. 3).

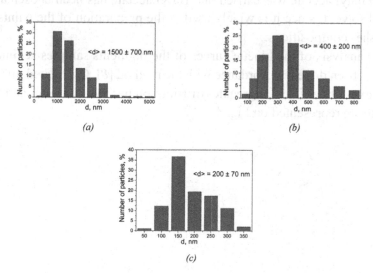

FIGURE 3 Zinc phosphate particles distribution upon size: *sample № 1 (a), sample № 2 (b), sample № 3 (c)*.

Less size of zinc phosphate nanoparticles obtained via their modification by *BMA*, in comparison with *BA*, probably can be explained both greatly higher evaporation heat of *BMA* (near 75 *calories/g* for *BMA* and 46 *calories/g* for *BA*), and some better solubility of *BA* in water (near 0.08% for *BMA* and 0.2% for *BA*) [9].

An elementary analysis of the obtained samples showed, that the ratio of the elements *Zn/P* is equal to 3/2 that confirms the formation exactly of zinc phosphate. At the same time, in *samples* 2 and 3 it was discovered, respectively, 1.1 and 2.3% of carbon; this points on the formation of polymeric film on the surface of nanoplates. Higher content of the carbon in *sample № 3* can be explained both by the increase of the specific surface of the pigment at the stabilization of zinc phosphate nanoplates by butylmethacrylate and by the formation of more compact polymeric film, than in a case of the stabilization with the use of *BA*.

11.3.3 SEDIMENTATION ANALYSIS OF ZINC PHOSPHATE NANOPLATES

In order to study the stability of zinc phosphate suspensions in different media the sedimentation analysis of the synthesized samples in water and in butyl acetate was carried out. Butyl acetate has been chosen as the model solvent, since it is widely used at the preparation of the paints and varnishes compositions.

An analysis of the kinetic curves of the pigments samples sedimentation has been done in accordance with the method [8]. Calculated differential curves of particles' aggregates distribution per sizes in water and butyl acetate are represented on Fig. 4.

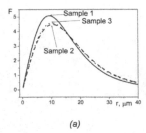

(a)

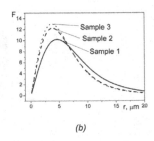

(b)

FIGURE 4 Distribution function *(F)* per radiuses of particles' aggregates for samples *№* 1–3 in water *(a)* and in butyl acetate *(b)*.

In spite of the fact that the used by us method of the sedimentation curves analysis is based on a range of the assumptions, carried out by us experiments permit to conclude that due to the modification by acrylic monomers the surface of the pigments became more hydrophobic. A significantly less average radius of the zinc phosphate particles' aggregates under sedimentation in organic solvent in comparison with their sedimentation in water confirms of this fact. At the same time, the obtained results of the sedimentation analysis are in good agreement with the results of the electron microscopy and the elementary analysis.

11.3.4 PROTECTIVE ACTION OF NANOSIZED ZINC PHOSPHATE IN ALKYD COATING ON THE ALUMINUM ALLOY

With the use of the electrochemical impedance spectroscopy the bilayer alkyd coatings based on the lacquer *PF*-170 superimposed on the duralumin alloy *D16T* was investigated. Such alloy has a much wide application in aircraft construction, transport, in building industry, etc. Typical protection scheme of the constructions in a case of such alloy application consists in the formation of the conversion layer on the surface of the metal, For example, at the expense of the oxidation, priming and upper varnish-and-paint layer.

In order to study the anticorrosive properties of the synthesized in the presence of *BMA* zinc phosphate nanoplates, the hinge of a pigment in quantity till 5% *mass* was added in priming layer of the alkyd coating *PF*-170 on the oxidized aluminum alloy. General thickness of the bilayer (inhibited coating + lacquer *PF*-170) coating was approximately 120 μm. In this same pentaphthalic coating the commercial zinc phosphate pigment *Novinox PZ02* was used for comparison. In order to discover the protective effect from the inhibited components the defects by diameter 1 mm through the coatings have been done. It was determined (Fig. 5), that the both commercial and synthesized by us nanosized zinc phosphate essentially increase the resistance of the charge transfer of aluminum alloy with the defect alkyd coating in corrosive medium. The interesting is fact, that the more anticorrosive effect is observed at the introduction into coating of the nanosized zinc phosphate in quantity from 1 till 3% *mass*, than 5% *mass*. Nanosized zinc phosphate is characterized by higher specific surface and that is why can possess by higher composition solubility and, correspondingly, will be characterized by better-inhibited properties. This indirectly means about the resistance decreasing for the corrosive solution contacting with the sample of coating having of zinc nanophosphate (Fig. 5). Alkyd coating with zinc nanophosphate at concentrations of 1 and 3% *mass* is characterized by the better anticorrosive protective properties than the coatings containing of 5% *mass* of the well-known zinc phosphate pigment *Novinox PZ02*. Such data point on the possibility to decrease the content of the inhibited pigment in alkyd coating in a case of the new synthesized by us nanosized pigment application on a basis of zinc phosphate.

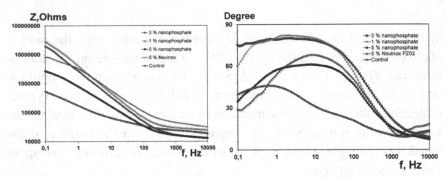

FIGURE 5   Frequency impedance dependencies of aluminum alloy *D16T* with inhibited alkyd coatings after 48 h in the medium of the acidic rain. Coatings contained the through defects by diameter 1 mm.

11.4   CONCLUSIONS

Nanosized inhibited pigments based on the zinc phosphate nanoparticles with the average linear sizes 200 *nm* and by the thickness 20 *nm* have been synthesized via ion change reaction in water-methanol medium. It was shown the possibility of acrylic monomers using for modification of the nanoparticles surface. It was determined, that the obtained nanoplates are characterized by better affinity to the organic solvent, in comparison with water. It was shown also, that the nanosized zinc phosphate is characterized by better-inhibited properties in comparison with the well-known phosphate pigment *Novinox PZ02* that permits to develop under industrial conditions the new competitive materials for their using in paint and varnish industry.

KEYWORDS

- acrylic monomers
- anticorrosion activity
- film- substrate
- metal-coating
- nanosized inhibited pigments
- zinc phosphate nanoplates

REFERENCES

1. Pokhmursky V., Kytsya A., Zin' I., Bazylyak L., Korniy S., Hrynda Yu. A method of obtaining of the nanosized zinc phosphate, *Patent of Ukraine* № 78529. Date of publ.: 25.03.2013. Bull. № 6. 1–4.
2. Pokhmursky V., Zin' I., Kytsya A., Bily L., Korniy S., Zin' Ya. Priming composition for anticorrosive coating, *Patent of Ukraine* № 78503. Date of publ.: 25.03.2013, Bull. № 6, 1–4.
3. Korsunsky L. F., Kalinskaya T. V., Stiopin S. N. Inorganic pigments, *Edition: "Khimiya"*. 1992, 336 p. *(in Russian)*
4. Rosenfeld I. L., Rubinshtein F. I., Zhygalova K. A. Protection of metals from the corrosion by paint and varnish coatings, M.: "Khimiya". 1987, 143 p. *(in Russian)*
5. Schuler D. Richtungsweisende Korrosionschutzpigmenten, *"Farbe und Lack"*. 1986, Bd. 92, № 8, 703–705 *(Application № 37317377 (Germany))*.
6. Zotov E. V., Lugantseva L. N., Petrov L. N. Protective properties of a series passivating pigments, *"Lakokrasochnyye materialy i ikh primienieniye"*. 1987, № 5, 27–29 *(in Russian)*.
7. Mazan P., Trojan M., Brandova D. *e. a.* Condensed Phosphates As Anticorrosive Pigments, *"Polymer Paint Colour Journal"*. 1990, v. 180, 605–606.
8. Tsiurupa N. N. Distribution of dispersed phase per size of the particles, *"Colloid Journal"*. 1964, v. 24, № 1, 117 – 125 *(in Russian)*.
9. Kargin V. A. Encyclopaedia of polymers. v. 1., M.: *"Sovietskaya encyklopedia"*. 1972, 1224 p. *(in Russian)*

CHAPTER 12

AN INFLUENCE OF KINETIC PARAMETERS OF REACTION ON THE SIZE OF OBTAINED NANOPARTICLES UNDER REDUCTION OF SILVER IONS BY HYDRAZINE

A. R. KYTSYA, L. I. BAZYLYAK, YU. M. HRYNDA, and YU. G. MEDVEDEVSKIKH

CONTENTS

SUMMARY

Spherical silver nanoparticles (NPs) were obtained by silver ions reduction with hydrazine in the presence of sodium citrate as a stabilizer. It has been investigated the kinetic regularities of the silver NPs nucleuses formation and their propagation depending on the starting concentration of the hydroxide ions and silver ions. It was investigated the influence of the synthesis conditions on the average diameter of the obtained silver NPs. It was shown the dependence of the obtained silver NPs on kinetic parameters of the process.

12.1 INTRODUCTION

One among the priority directions of science and technology development is the creation of nanomaterials having the unique physical, electrochemical and catalytic properties [1–5], that gives the wider possibilities for development of the new effective catalysts, sensory systems, preparations with a high biological activity. Colloidal nanoparticles of the metals possessing by special properties differing from the isolated atoms and from the mass metal call an especial attention.

There are a number of methods for the synthesis of different upon nature nanoparticles and nanomaterials, however the kinetic peculiarities and regularities of the formation (nucleation and propagation) of nanoparticles studied insufficiently.

The aim of our work was to investigate the kinetic regularities of the silver nanoparticles synthesis via reduction reaction of silver nitrate by hydrazine in the presence of the sodium citrate depending on the concentration of the hydroxide ions and of the silver ions.

12.2 EXPERIMENTAL PART

Synthesis of the silver nanoparticles was carried out in the glass reactor with the thermostated shell equipped with the magnetic stirrer and thermometer at 20°C via the reduction reaction of the silver nitrate with hydrazine in aqueous medium in the presence of the sodium hydroxide. Sodium citrate was used as the stabilizer of silver nanoparticles.

Kinetic of the reaction was studied by potentiometry with the use of the ion-selective microelectrode *ELIS*-131 Silver. The concentration of the silver ions was determined continuous during the reaction proceeding per change of the potential of the ion-selective electrode regarding to the chlorine-silver comparison electrode.

The form and the average diameter of the silver nanoparticles were estimated with the use of the scanning electron microscopy EV*O-40XVP (Carl Zeiss)* with a system of the *X*-ray microanalysis *INCA* Energy, *XRD*-analysis and also on a basis of the adsorption spectrum of the surface Plasmon resonance [5–7].

XRD-analysis was carried out with the use of *XRD* diffractometer *DRON*-3.0 with Cu-K$_a$ irradiation ($\lambda = 1.5405$ *nm*). Data have been analyzed by full-profile revision in accordance with the *Ritveld's* method with the use of the simulation package *GSAS (General Structure Analysis System)*.

Investigations of the adsorption spectrum of surface Plasmon resonance of the silver nanoparticles sols were carried out with the use of the single-beam spectrophotometer UV-visible range *UVmini*-1240 (*P/N* 206-89175-92; *P/N* 206-89175-38; *Shimadzu Corp., Kyoto, Japan*).

12.3 RESULTS AND DISCUSSION

In order to explain and to investigate the chemical process of the silver nanoparticles synthesis the kinetic characteristics have been investigated, namely the change of the concentration of the silver ions during the experiment.

Typical kinetic curves of the silver concentration change are represented on Fig. 1.

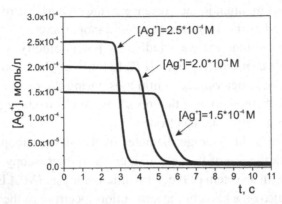

FIGURE 1   Kinetic curves of the reduction reaction of silver ions by hydrazine at different starting concentration's of the silver nitrate.

As we can see, the starting section of the kinetic curve corresponds to the stage of the nucleus centers formation and the following sharp decrease of the concentration of silver ions corresponds to their growth stage.

The rates of the nucleus centers formation ($W0 = 1/t_0$) of silver nanoparticles were calculated on the basis of the time length (t_0) of the starting section of the kinetic curve.

At the analysis of the experimental data it was discovered that the rate of the nanoparticles nucleation (W_0) linearly depends on the concentrations of sodium hydroxide and on the concentration of the silver ions (Fig. 2 a, b). Such dependencies confirm the first reaction order for the silver nanoparticles nucleation reaction per hydroxide ions and per silver ions.

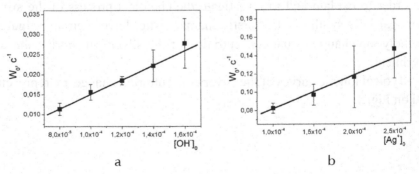

FIGURE 2   Dependence of the silver nanoparticles nucleation rate on the starting concentration of sodium hydroxide (a) and silver nitrate (b).

Since the rate of the silver ions reduction will be maximal in the point of inflection of the kinetic curve, then on a basis of the experimental data (Fig. 1) it were found the values of $AgNO_3$ concentration in this point ($[Ag^+]_{max}$). Accordingly to the reaction equation the values of hydroxide-ions concentration ($[OH^-]_{max}$) in point, corresponding to the maximal process rate have been calculated. The values of maximal rate of the silver nanoparticles growth (W_{max}) was estimated on a basis of the tangent line inclination to the kinetic curve in the point of inflection.

$$4\ AgNO3 + 4\ NaOH + N2H4 = 4\ Ag + 4\ NaNO3 + 4\ H2O + N2\uparrow$$

It was determined that the rate of the silver nanoparticles growth can be also described by the linear dependence on concentration of ions hydroxide and silver nitrate (Fig. 3, a, b).

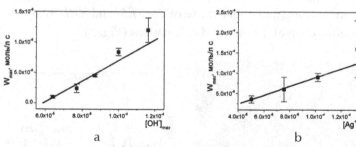

a   b

FIGURE 3 Dependence of the silver nanoparticles growth rate on concentration of sodium hydroxide (a) and silver nitrate (b).

In order to identify the obtained silver nanoparticles, their spectral characteristics were investigated (Fig. 4, a). The spectrum of silver nanoparticles adsorption is characterized by one maximum corresponding to their spherical form. Analyzing the references [5–7], it was discovered that the value of the square of wave frequency in adsorption maximum of the surface Plasmon resonance of silver nanoparticles linearly depends on their size (Fig. 4, b), that gives the possibilities to calculate an average diameter of the obtained silver nanoparticles. Calculated values of the average diameter of silver nanoparticles consist of 12 – 35 nm.

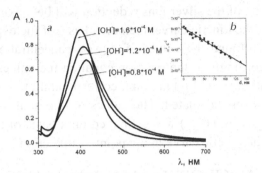

FIGURE 4 Adsorption electron spectra of silver nanoparticles obtained at different starting concentrations of sodium hydroxide (a) and calibration plot for calculation of their average diameter (b) based on the references [5–7].

Silver nanoparticles obtained at 20 °C and starting concentrations of reagents $[AgNO_3]_0 = 2.5 \times 10^{-4}$ M, $[NaOH]_0 = 3.0 \times 10^{-4}$ M, $[N_2H_4]_0 = 7.5 \times 10^{-5}$ M were investigated with the use of the *SEM* and *XRD* analysis methods for confirmation of the obtained calculations (Fig. 5).

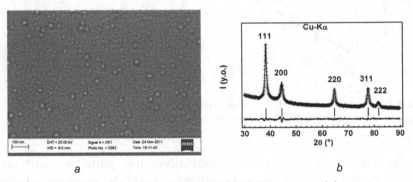

FIGURE 5 *SEM* image (a) and *XRD*-spectrum (b) of silver nanoparticles.

On a basis of the *XRD*-analysis results, an average size of the silver crystallites was calculated; it consists of $D_V = 9.3$ *nm*; respectively the diameter of the spherical particle for monodisperse system is $D = 4/3$ and $D_V = 12.4$ *nm*. Calculated accordingly to the location of the surface Plasmon resonance adsorption maximum of the sol silver nanoparticles (Fig. 4, *b*) value of average diameter of silver nanoparticles obtained under such conditions consists of 12 nm.

At the analysis of the experimental data it was found that the size of the obtained silver nanoparticles depends on the ratio of the nuclear centers formation rate and the nuclear centers growth rate (Fig. 6).

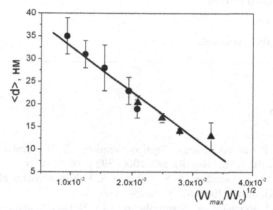

FIGURE 6 Dependence of an average diameter of silver nanoparticles on kinetic parameters of the process: • – an average diameter of silver nanoparticles obtained at different starting concentrations of sodium hydroxide; ▲ – an average diameter of silver nanoparticles obtained at different starting concentrations of $AgNO_3$.

Evidently, such dependence can be explained by fact that at the nuclear centers formation rate increasing not only the concentration of the nuclear centers is increased, but also their critical radius is decreased that leads to the decreasing of the average size of synthesized nanoparticles.

12.4 CONCLUSIONS

1. It was determined that the rates of the nuclear centers formation and the nuclear centers growth are linearly depend on concentrations of ions hydroxides and silver ions.
2. It was established the linear dependence of average diameter of the obtained silver nanoparticles on the ratio of the nuclear centers formation rate and the nuclear centers growth rate in exponent ½.

KEYWORDS

- hydrazine
- kinetic parameters
- reaction rate
- silver nanoparticles

REFERENCES

1. Suzdalyev I. P. Nanotechnology: Physico-chemistry of Nanoclasters, Nanostructures and Nanomaterials, M.: "KomKniga", 2006, 592 p. *(in Russian)*.
2. Pomogaylo A. D., Rosenberg A. S., Uflyand A. S. Nanoparticles of Metals in Polymers, M.: "Khimiya", 2002, 672 p. *(in Russian)*.
3. Egorova E. M., Revina A. A., Rostovshchikova T. N. Bactericidal and Catalytic Properties of Stable Metallic Nanoparticles in the Reverse Micelles, *"Vestnik Moskovskogo Universitieta" Ser. 2. "Khimiya"*. 2001, v. 42, № 5, 332–338 *(in Russian)*.
4. Tierskaya I. N., Salnikov D. S., Makarov S. V. and etc. Chemical Synthesis of Stable Nanosized Water-Organic Disperse Copper, *"Physico Khimiya Povierkhnosti I Zashchita Materialov"*. 2008, v. 44, № 5, 503–505 *(in Russian)*.
5. Krutyakov Yu. A., Kudrinsky A. A., Olenin A. Yu. Synthesis and Properties of Silver Nano-particles: Achievements and Perspectives, *"Uspiekhi Khimiji"* **77 (3) 2008 C, 242–269** *(in Russian)*.
6. David D. Evanoff Jr., Chumanov G. Synthesis and Optical Properties of Silver Nanoparticles and Arrays, *"Chem. Phys. Chem."*. 2005, v. 6, 1221–1231.
7. Kryukov A. I., Zinchuk N. N., Korzhak A. V. The Influence of Conditions of Catalytic Synthesis of Metallic Silver Nanoparticles on Their Plasmon Resonance, *"Theoret. i Experiment. Khimiya"*. 2003, 39, № 1, 8–13 *(in Russian)*.

CHAPTER 13

CASE STUDIES ON NANOPOLYMERS AND THEIR CHEMICALS COMPLEXITY—NEW APPROACHES, LIMITATIONS AND CONTROL

A. K. HAGHI and G. E. ZAIKOV

CONTENTS

SUMMARY

The field of nanotechnology is one of the most popular areas for current research and development in basically all technical disciplines. This chapter presents five case studies on nanopolymers and their chemicals complexities.

For the first case study, the particulate filled nanopolymers is studied. An investigation on viscometric flow for particulate filled nanopolymers is presented as the second case study in this chapter. Application of synthetic or natural inorganic fillers is reviewed as the third case study. The next two case studies are devoted to description of a multiscale micromechanical model and application of cement materials reinforcement with nanoparticles.

13.1   INTRODUCTION

Polymer nanocomposites, a class of polymers reinforced with low quantities of well-dispersed nanoparticles offer advantages over conventional composites. It has been proved that when the sizes of nanofillers are very small (at least one of their dimensions is under 100 nm), the interface regions are so large that they start to interact at very low level of loadings. Nanoparticles have great effect on the properties and morphology of polymeric nanocomposites due to their large specific surface and high surface energy [1–10] The interactions between polymer matrix and nanoparticles alter polymer chemistry, that is, chain mobility and degree of cure and generate new trap centers in the composite which brings about the significant change in electrical properties. Where the size of the particle is close to that of the polymer chain length, the particles do not behave like foreign inclusions and space charge densities are small [1–15]. In this chapter, five case studies on different types of nanopolymers are presented.

13.2   CASE STUDY I—PARTICULATE FILLED NANOPOLYMERS

Particulate filled polymers are used in very large quantities in all kinds of applications. In spite of the overwhelming interest in nanocomposites, biomaterials and natural fiber reinforced composites, considerable research and development is done on particulate filled polymers even today. The reason for the continuing interest in traditional composites lies, among others, in the changed role of particulate fillers. In the early days fillers were added to the polymer to decrease the price. However, the ever-increasing technical and esthetic requirements as well as soaring material and compounding costs require the utilization of all the possible advantages of fillers. Fillers increase stiffness and heat deflection temperature, decrease shrinkage and improve the appearance of the composites. Productivity can be also increased in most processing technologies due to their decreased specific heat and increased heat conductivity. Fillers are very often introduced into the polymer to create new functional properties not possessed by the matrix polymer at all, such as flame retardancy or conductivity.

The properties of all heterogeneous polymer systems are determined by the same four factors: component properties, composition, structure and interfacial interactions.

Although certain fillers and reinforcements including layered silicates, other nanofillers, or natural fibers possess special characteristics, the effect of these four factors is universal and valid for all particulate filled materials.

The modern methods of experimental and theoretical analysis of polymer material structure and properties allow not only the confirmation of earlier propounded hypotheses, but the obtaining of new results as well. This chapter considers some important problems of particulate-filled polymer nanocomposites, the solution of which advances substantially the understanding and prediction of these materials' properties. Polymer nanocomposites multicomponentness (multiphaseness) requires the determination of their structural components quantitative characteristics. In this aspect interfacial regions play a particular role, as a reinforcing element. Therefore, the knowledge of interfacial layer dimensional characteristics is necessary for quantitative determination of one of the most important parameters of polymer composites in general (their degree of reinforcement) [10–27].

The aggregation of the initial nanofiller powder particles in more or less large particle aggregates always occurs in the course of the technological process of making particulate-filled polymer composites in general and elastomeric nanocomposites in particular. The aggregation process explains macroscopic properties of the nanocomposites.

Nowadays, nanofiller aggregation process in nanocomposites has gained special attention. In general, nanofiller particles aggregate size exceeds 100 nm – the value, which is assumed (though conditionally enough) as an upper dimensional limit for nanoparticles. Therefore, at present several methods exist, which allow the suppression of the nanoparticles aggregation process. This also assumes the necessity of the nanoparticles aggregation process in quantitative analysis [15–37].

It is well known, that in particulate-filled elastomeric nanocomposites (rubbers), nanofiller particles form linear spatial structures ('chains'). At the same time in polymer composites, filled with dispersed microparticles (microcomposites), particles (aggregates of particles) of filler form a fractal network, which defines the polymer matrix structure (analog of fractal lattice in computer simulation). This results in different mechanisms of polymer matrix structure formation in micro and nanocomposites. If in the first filler, the particles' (aggregates of particles) fractal network availability results in a 'disturbance' of the polymer matrix structure (that is expressed in the increase of its fractal dimension d_f) then for polymer nanocomposites as the nanofiller contents change the value d_f is not changed and is equal to the matrix polymer structure fractal dimension. As it has been expected, the change of the composites of the indicated classes structure formation mechanism change defines their properties, in particular, the degree of reinforcement. Therefore, nanofiller structure fractality is very important to study.

However, the scale effects in composites should be well known. The dependence of failure stress on grain size for metals or of effective filling degree on filler particles size for polymer composites are examples of such an effect. The strong dependence of elasticity modulus on nanofiller particles' diameter is observed for particulate-filled elastomeric nanocomposites. Therefore it is necessary to elucidate the physical grounds of nano and micromechanical behavior scale effect for polymer nanocomposites.

At present a wide number of dispersive materials are known, which are able to strengthen elastomeric polymer materials. These materials are very diverse in their surface chemical constitution, but small size particles are a common feature for them. On the basis of this observation the hypothesis was offered, that any solid material would strengthen the rubber at the condition that it was in a much-dispersed state and it could be dispersed in polymer matrix. It should be pointed-out, that filler particles of a small size are necessary and, probably, the main requirement for reinforcement effect realization in rubbers. Using modern terminology, one can say, that for a rubber's reinforcement the nanofiller particles (for which their aggregation process is suppressed as far as possible) would be the most effective ones. Therefore, the theoretical analysis of a nanofiller particle's size influence on polymer nanocomposites reinforcement is necessary.

The purpose of the present chapter is to describe the solution of the previously described problems with the help of modern experimental and theoretical techniques for the example of particulate-filled butadiene-styrene rubber (BSR).

13.2.1   EXPERIMENTAL

BSR, which contains 7.0–12.3% *cis*- and 71.8–72.0% *trans*-bonds, with a density of 920–930 kg/m^3 was used as a matrix polymer. This rubber is a fully amorphous one.

Fullerene-containing mineral shungite of Zazhoginsk's deposit consists of ~30% globular amorphous metastable carbon and ~ 70% highly dispersed silicate particles. Industrially made, technical carbon (TC) was used as a nanofiller. The TC, nano- and microshungite particles average size is 20, 40 and 200 nm, respectively. The indicated filler content is equal to 37 mass%. Nano and microdimensional dispersed shungite particles were prepared from industrially output material from the original technology processing. The size and polydispersity analysis of the shungite particles in the milling process was monitored with the aid of an analytical disk centrifuge (allowing the determination, with high precision, of the size and distribution by the size within limited range).

Nanostructure was studied using atomic force microscopes (AFM). AFM results were processed with the help of a specialized software package scanning probe image processor (SPIP). SPIP is a powerful programme package for processing images, obtained from scanning probe microscopy (SPM), AFM, scanning probe micrsocopy (STM), scanning electron microscopes, transmission electron microscopes, interferometers, profilometers, optical microscopes and so on. The software possesses the whole number function, which is necessary for precise image analysis, including:

- The possibility of obtaining three-dimensional reflecting objects.
- Quantitative analysis of particles or grains – more than 40 parameters can be calculated for each particle or pore found: area, perimeter, mean diameter, the ratio of linear sizes of grain width to its height, distance between grains, co-ordinates of grain center of mass can be presented in a diagrammatic form or as a histogram form.

The tests on elastomeric nanocomposites to determine their nanomechanical properties were carried out by a nano-indentation method (in loads of a wide range from 0.01 mN up to 2.0 mN). Sample indentation was conducted at 10 points with defined intervals. The load was increased with constant rate up to the greatest given load). The indentation rate was changed in conformity with the greatest load value counting – that loading cycle should take 20 seconds. The unloading was conducted at the same rate as the loading. In this experiment the 'Berkovich indentor' was used with the angle at the top of 65.3° and a rounding radius of 200 nm. Indentations were carried out in the checked load regime with a preload of 0.001 mN.

For elasticity modulus calculation (obtained in the experiment by nano-indentation), coarse dependences of load on indentation depth (strain) at ten points for each sample at loads of 0.01, 0.02, 0.03, 0.05, 0.10, 0.50, 1.0 and 2.0 mN were processed according to the Oliver-Pharr method.

13.2.2 RESULTS AND DISCUSSION

Figure 1.1 represents the elasticity moduli calculation for nanocomposite BSR/nanoshungite components (matrix, nanofiller particle and interfacial layers), derived from the interpolation process of the nano-indentation data. The processed SPIP polymer nanocomposite image with shungite nanoparticles allows experimental determination of the interfacial layer thickness (l_{if}), as steps on elastomeric matrix-nanofiller boundary. The measurements of 34 such steps (interfacial layers) width on the processed in SPIP images of the interfacial layer at various sections gave a mean experimental value of l_{if} = 8.7 nm. The nano-indentation results showed that the elasticity modulus of the interfacial layers was only 23–45% lower than the elasticity modulus of the nanofiller (but it was higher than the corresponding parameter of the polymer matrix by 6.0–8.5 times). These experimental data confirm that for the nanocomposite studied, the interfacial layer is actually a reinforcing element to the same extent, as the nanofiller.

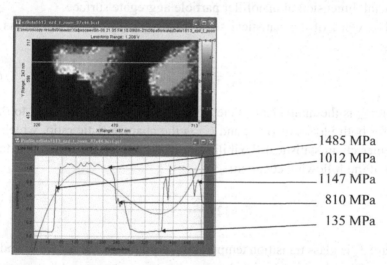

FIGURE 1.1 The processed by SPIP image of nanocomposite BSR/nanoshungite, obtained by the force modulation method, and mechanical characteristics of the structural components according to the data of nano-indentation (strain 150 nm).

Let us consider further the theoretical estimation of the value, l_{if}, according to the two methods and compare these results with the ones obtained experimentally. The first method simulates the interfacial layer in the polymer composites as a result of the interaction of two fractals – the polymer matrix and the nanofiller surface. In this case there is a sole linear scale l, which defines the interpenetration distance of the fractals. Since the nanofiller elasticity modulus is essentially higher than the corresponding parameter for rubber (in this case – by 11 times, see Fig. 1.1), then the indicated interaction reduces to nanofiller indentation in the polymer matrix and then $l = l_{if}$.

In this case it can be written:

$$l_{if} \approx a\left(\frac{R_p}{a}\right)^{2(d-d_{surf})/d} \tag{1.1}$$

where: a is a lower linear scale of fractal behavior, which is accepted for polymers as being equal to the statistical segment length (l_{st}); R_p is a nanofiller particle (more precisely, particle aggregates) radius, which for nanoshungite is equal to ~84 nm; d is dimension of Euclidean space, in which the fractal is considered (it is obvious, that in our case $d = 3$); d_{surf} is a fractal dimension of nanofiller particle aggregate surface.

The value of the statistical segment length (l_{st}) is determined as follows:

$$l_{st} = l_0 C_\infty \tag{1.2}$$

where: l_0 is the main chain skeletal bond length, which is equal to 0.154 nm for both blocks of BSR; and C_∞ is the characteristic ratio, which is a polymer chain statistical flexibility indicator, and is determined with the help of the following equation:

$$T_g = 129\left(\frac{S}{C_\infty}\right)^{1/2} \tag{1.3}$$

where: T_g is glass transition temperature, equal to 217 K for BSR; and S is macromolecule cross-sectional area.

As is already known, the macromolecule diameter quadrate values are equal: for polybutadiene (PB) – 20.7 Å^2 and for polystyrene (PS) – 69.8 Å^2. Having calculated the cross-sectional area of the macromolecule, simulated as a cylinder, for the indicated polymers according to the known geometrical formulas, we obtain diameters of 16.2 Å^2 and 54.8 Å^2, for PB and PS, respectively. Furthermore, accepting as S the average value of the cross-sectional areas, we obtain for BSR $S = 35.5$ Å^2. Then according to Eq. (1.3) at the indicated values of T_g and S we obtain $C_\infty = 12.5$ and according to Eq. (1.2), $l_{st} = 1.932$ nm.

The fractal dimension of the nanofiller surface (d_{surf}) was determined with the help of Eq. (1.4):

$$S_u = 410 R_p^{d_{surf} - d} \qquad (1.4)$$

where: S_u is a nanoshungite particle specific surface, calculated as follows:

$$S_u = \frac{3}{\rho_n R_p} \qquad (1.5)$$

where: ρ_n is the nanofiller particles aggregate density, determined according to the following formula:

$$\rho_n = 0.188 \left(R_p \right)^{1/3} \qquad (1.6)$$

The calculation according to Eqs. (1.4–1.6) gives $d_{surf} = 2.44$. Furthermore, using these equations and related parameters, from the Eq. (1.1) the theoretical value of interfacial layer thickness = 7.8 nm. This value is close enough to the one obtained experimentally (their discrepancy makes up ~10%).

The second method of value estimation consists in using the following two equations:

$$\varphi_{if} = \varphi_n \left(d_{surf} - 2 \right) \qquad (1.7)$$

and

$$\varphi_{if} = \varphi_n \left[\left(\frac{R_p + l_{if}^T}{R_p} \right)^3 - 1 \right] \qquad (1.8)$$

where: φ_{if} and φ_n are relative volume fractions of interfacial regions and nanofiller, respectively.

The combination of the indicated equations gives the following formula for calculation:

$$l_{if}^T = R_p \left[\left(d_{surf} - 1 \right)^{1/3} - 1 \right] \qquad (1.9)$$

The calculation according to Eq. (1.9) gives for the nanocomposite considered = 10.8 nm, which also corresponds well enough to the experiment (in this case the discrepancy between l_{if} and is ~19%).

Let us note in conclusion an important experimental observation, which follows from the results of the surface scan of nanocomposite studied and processed by SPIP (Fig. 1.1).

As one can see, at one nanoshungite particle surface from one to three (on average, two) steps can be observed, structurally identified as interfacial layers. It is significant that in these steps, the step width (or l_{if}) is approximately equal to the first (the closest to the nanoparticle surface) step width. Therefore, the indicated observation supposes, that in elastomeric nanocomposites on average two interfacial layers are formed: the first, at the nanofiller particle surface with the elastomeric matrix interaction, as a result of which molecular mobility in this layer is frozen and its state is a glassy-like one, and the second, at the glassy interfacial layer with the elastomeric polymer matrix interaction. From the practical point of view the most important question is whether one interfacial layer or both serve as nanocomposite reinforcing elements. Let us consider the following quantitative estimation. The degree of reinforcement (E_n/E_m) of polymer nanocomposites is given by Eq. (1.10):

$$\frac{E_n}{E_m} = 1 + 11\left(\varphi_n + \varphi_{if}\right)^{1.7} \tag{1.10}$$

where: E_n and E_m are elasticity moduli of the nanocomposite and the matrix polymer, respectively ($E_m = 1.82$ MPa).

According to the Eq. (1.7) the sum ($\varphi_n + \varphi_{if}$) is equal to:

$$\varphi_n + \varphi_{if} = \varphi_n\left(d_{surf} - 1\right) \tag{1.11}$$

if one interfacial layer (the closest to the nanoshungite surface) is a reinforcing element and:

$$\varphi_n + 2\varphi_{if} = \varphi_n\left(2d_{surf} - 3\right) \tag{1.12}$$

if both interfacial layers are a reinforcing element.

In its turn, the value φ_n is determined according to Eq. (1.13):

$$\varphi_n = \frac{W_n}{\rho_n} \tag{1.13}$$

where: W_n is nanofiller mass content, ρ_n is its density, determined according to Eq. (1.6).

The calculation according to Eqs. (1.11) and (1.12) gave the following E_n/E_m values: 4.60 and 6.65, respectively. Since the experimental value $E_n/E_m = 6.10$ is closer to the value, calculated according to Eq. (1.12), then this means that both interfacial layers are a reinforcing element for the nanocomposites studied. Note: Eq. (1.1) in its initial form was obtained as a relationship with a proportionality sign, that is, without fixed proportionality coefficient.

Thus, the previously used nanoscopic method allows the estimation of both special features in the interfacial layer structure in polymer nanocomposites and, its sizes and properties. For the first time it has been shown, that in elastomeric particulate-filled nanocomposites two consecutive interfacial layers are formed, which are a reinforcing element for the indicated nanocomposites. The proposed theoretical method of interfacial layer thickness estimation, elaborated within the frameworks of fractal analysis in experimental works.

For theoretical treatment of nanofiller particles aggregate growth processes and final sizes, traditional irreversible aggregation models are inapplicable, since it is obvious, that in nanocomposite aggregates a large number of simultaneous growth takes place. Therefore, the model of multiple growths was used for nanofiller aggregation description.

In Fig. 1.2, the images of the nanocomposites studied, obtained in the force modulation regime, and corresponding to them nanoparticles aggregate, fractal dimension (d_f) distributions are shown. As it follows from the deduced values of $(d_f^{ag} = 2.40–2.48)$, nanofiller particles aggregates in the studied nanocomposites are formed by a mechanism of particle-clustering (P-Cl), that is, they are Witten-Sander clusters. The variant A, was chosen, which according to mobile particles are added to the lattice, consisting of a large number of 'seeds' with a density of c_0 at the beginning of the simulation. Such a model generates the structures, which have a fractal geometry of short scale length with a value of $d_f \gg 2.5$ (see Fig. 1.2) and homogeneous structure of large length scales. A relatively high particle concentration (c) is required in the model for uninterrupted network formation.

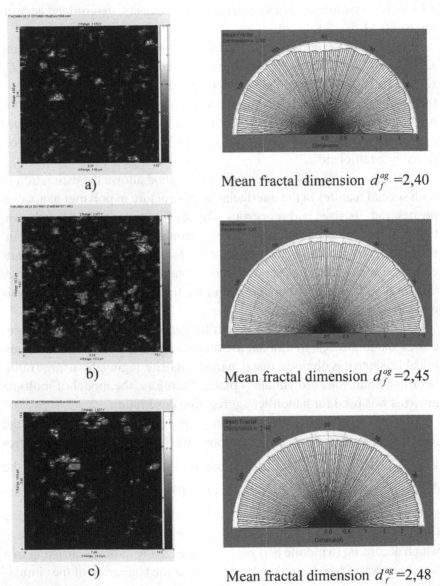

a)

Mean fractal dimension $d_f^{ag} = 2,40$

b)

Mean fractal dimension $d_f^{ag} = 2,45$

c)

Mean fractal dimension $d_f^{ag} = 2,48$

FIGURE 1.2 The images, obtained in the force modulation regime, for nanocomposites, filled with: (a) TC (b) nanoshungite, and (c) microshungite, and corresponding to them fractal dimensions d_f^{ag}.

For high concentration of c_0 following relationship can be obtained:

$$R_{max}^{d_f^{ag}} = N = c/c_0 \qquad (1.14)$$

where: R_{max} is nanoparticles cluster (aggregate) greatest radius, N is nanoparticles number per aggregate, c is nanoparticles concentration, c_0 is 'seeds' number, which is equal to the number of nanoparticle clusters (aggregates).

The value N can be estimated according to Eq. (1.15):

$$2R_{max} = \left(\frac{S_n N}{\pi\eta}\right)^{1/2} \qquad (1.15)$$

where: S_n is cross-sectional area of the nanoparticles of which an aggregate consists; and h is a packing coefficient, equal to 0.74.

The experimentally obtained nanoparticles aggregate diameter $2R_{ag}$ was accepted as $2R_{max}$ (Table 1.1) and the value S_n was also calculated according to the experimental values of the nanoparticle radius, r_n (Table 1.1).

TABLE 1.1 The parameters of the irreversible aggregation model of nanofiller particles aggregate growth.

Nanofiller	R_{ag} (nm)	r_n (nm)	N	R_{max}^T (nm)	R_g^T (nm)	R_c (nm)
TC	34.6	10	35.4	34.7	34.7	33.9
Nanoshungite	83.6	20	51.8	45.0	90.0	71.0
Microshungite	117.1	100	4.1	15.8	158.0	255.0

In Table 1.1 are shown the values of N for the nanofillers studied, obtained according to the indicated method. It is significant that the value N is a maximum one for nanoshungite despite it having larger values for r_n in comparison with TC.

Equation (1.14) allows the greatest radius of nanoparticles aggregate within the frameworks of the aggregation model to be estimated. These values of are shown in Table 1.1, from which their reduction in a sequence of TC, nanoshungite, microshungite, fully contradicts the experimental data, that is, to change of R_{ag} (Table 1.1). However, we must not neglect the fact that Eq. (1.14) was obtained from a computer simulation, where the initial aggregating particles size are the same in all cases. For real nanocomposites the values of r_n can be distinguished (Table 1.1). It is expected, that the value R_{ag} or will be the higher, the larger the radius of the nanoparticles forming the aggregate, that is, r_n. Then the theoretical value of the nanofiller particles cluster (aggregate) radius can be determined as follows:

$$R_{ag}^T = k_n r_n N^{1/d_f^{ag}}$$
(1.16)

where: k_n is proportionality coefficient, in the present work accepted empirically as equal to 0.9.

The comparison of the experimental R_{ag} and values of the nanofiller particles aggregates radius (calculated according to Eq. (1.16)) shows their good agreement (the average discrepancy of R_{ag} and is 11.4 %). Therefore, the theoretical model shows good agreement to the experimental results only for the real characteristics of the aggregating particles and, in the first place, their size.

Let us consider two more important aspects of nanofiller particles aggregation within the frameworks of the model. Some features of the indicated process are defined by nanoparticle diffusion during nanocomposites processing. Specifically, the length, connected with a diffusible nanoparticle, is correlated to the length of diffusion x. By definition, the growth phenomena in sites, remote more than x, are statistically independent. Such definition allows the value x to be connected with the mean distance between the nanofiller particle aggregates L_n. The value x can be calculated according to the equation:

$$\xi^2 \approx c^{-1} R_{ag}^{d_f^{ag}-d+2}$$
(1.17)

where: c is the concentration of nanoparticles, which should be accepted to be equal to the nanofiller volume contents φ_n, which is calculated according to the Eqs. (1.6) and (1.13).

The values r_n and R_{ag} were obtained experimentally (see the histogram of Fig. 1.3). In Fig. 1.4 the relation between L_n and x is shown, which, as it is expected, proves to be linear and passing through coordinates origin. This means, that the distance between nanofiller particle aggregates is limited by mean displacement of statistical spaces by which nanoparticles are simulated. The relationship between L_n and x can be expressed analytically as:

$$L_n \approx 9.6\xi, \text{ nm} \tag{1.18}$$

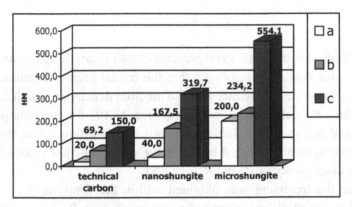

FIGURE 1.3   (a) The initial particles diameter (b) their aggregate size in nanocomposite and (c) distance between the nanoparticle aggregates for nanocomposites, filled with TC, nanoshungite and microshungite.

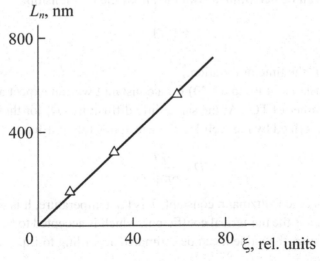

FIGURE 1.4   The relation between diffusion correlation length (x) and distance between nanoparticle aggregates (L_n) for the nanocomposites considered.

The second important aspect of the model in reference to nanofiller particle aggregation simulation is a finite nonzero initial particles concentration c or φ_n effect, which takes place in any real systems. This effect is realized at the condition $\xi \approx R_{ag}$, that occurs at the critical value $R_{ag}(R_c)$, determined according to the following relationship:

$$c \sim R_c^{d_f^{ag}-d} \qquad (1.19)$$

The right side of Eq. (1.19) represents cluster (particle aggregate) mean density. This equation establishes that the fractal growth continues only, until the cluster density reduces up to a medium density, in which it grows. The values R_c, calculated according to Eq. (1.19) for the nanoparticles considered are shown in Table 1.1, from which follows, that they give reasonable agreement with this parameter's experimental values R_{ag} (the average discrepancy of R_c and R_{ag} is 24%).

Since the treatment was obtained within the framework of a more general model of diffusion-limited aggregation, then its agreement to the experimental data indicated unequivocally that aggregation processes in these systems were controlled by diffusion. Therefore, let us consider briefly nanofiller particle diffusion. Statistically, Walker's diffusion constant (z) can be determined with the aid of the relationship:

$$\xi \approx (\zeta t)^{1/2} \qquad (1.20)$$

where: t is the time or duration.

For instance if in Eq. (1.20) for a constant t we can expect a z increase in the number of TC). At the same time diffusivity (D) for these particles can be described by the well-known Einstein's relationship:

$$D = \frac{kT}{6\pi\eta r_n \alpha} \qquad (1.21)$$

where: k is the Boltzmann constant, T is the temperature, h is the medium viscosity, a is the numerical coefficient, which is accepted to be equal to 1.

In its turn, the value h can be estimated according to Eq. (1.22):

$$\frac{\eta}{\eta_0} = 1 + \frac{2.5\varphi_n}{1-\varphi_n} \qquad (1.22)$$

where: η_0 and h are initial polymer and its mixture with nanofiller viscosity, respectively.

The calculation using Eqs. (1.21) and (1.22) shows, that within the indicated nanofiller number, the value D changes from 1.32 to 1.14 to 0.44 relative units (i.e., reduces each time, which was expected). This apparent contradiction is due to the choice of the condition t = constant (where t is the nanocomposite production duration) in Eq. (1.20). Considering real conditions the value t is restricted by the nanoparticle contact with growing aggregate and then instead of t the value t/c_0 should be used, where c_0 is the seed concentration, determined according to Eq. (1.14). In this case the value z for the indicated nanofillers changes from 0.288 to 0.118 to 0.086 elative units (i.e., it reduces in 3.3 times) which agrees fully with the calculation using Einstein's relationship (Eq. 1.21). This means, that nanoparticle diffusion in the polymer matrix obey the classical laws of Newtonian rheology.

Thus, the dispersed nanofiller particles aggregation in elastomeric matrix can be described theoretically within the frameworks of a modified model of irreversible aggregation particle-cluster. The obligatory consideration of nanofiller initial particle size is a feature of the model's application to real systems. The particles' diffusion in the polymer matrix obeys classical laws of Newtonian liquid hydrodynamics. This approach allows prediction of the nanoparticle aggregates final parameters as a function of the initial particle size, their content and other factors.

At present there are several methods of filler structure (distribution) determination in the polymer matrix, both experimental and theoretical. All the indicated methods describe this distribution by a fractal dimension D_n of the filler particle network. However, the correct determination of any object fractal (Hausdorff) dimension includes three obligatory conditions. The first of them is the previous determination of fractal dimension numerical magnitude, which should not be equal to the object topological dimension. As it is known, any real (physical) fractal possesses fractal properties within a certain scales. Therefore, the second condition is the evidence of object self-similarity in this range. And the third condition is the correct choice of measurement scales itself. The first method of dimension D_n experimental determination uses the following fractal relationship, Eq. (1.23):

$$D_n = \frac{\ln N}{\ln \rho} \qquad (1.23)$$

where: N is the number of particles with size r.

Particles sizes were established using atomic power microscopy data (see Fig. 1.2). For each of the three nanocomposites studied not less than 200 particles were measured (the sizes of which were divided into 10 groups and the mean values of N and r were obtained). The dependences $N(r)$ in double logarithmic coordinates were plotted, which proved to be linear and the values of D_n were calculated according to their slope (see Fig. 1.5). It is obvious, that from such an approach that the fractal dimension D_n is determined in two-dimensional Euclidean space, whereas real nanocomposites should be considered in three-dimensional Euclidean space. The following relationship can be used for D_n recalculation for a three-dimensional space:

$$D3 = \frac{d + D2 \pm \left[(d - D2)^2 - 2\right]^{1/2}}{2} \qquad (1.24)$$

where: $D3$ and $D2$ are corresponding fractal dimensions in three- and two-dimensional Euclidean spaces, D is dimension of Euclidean space.

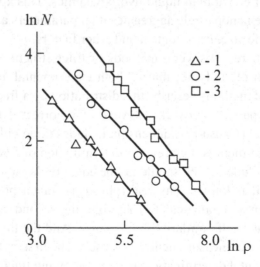

FIGURE 1.5 The dependencies of nanofiller particles number N on their size r for nanocomposites BSR/TC (1), BSR/nanoshungite (2) and BSR/microshungite (3).

The method dimensions D_n calculated are shown in Table 1.2. As is shown from the data of this table, the values of D_n for the nanocomposites studied vary within the range of 1.10–1.36 (i.e., they are more or less a branched linear formation of nanofiller particles in an elastomeric nano-composite structure).

It should be remembered that for particulate-filled composites poly-hydroxyether/graphite the value D_n changes within the range of approx. 2.30–2.80 (i.e., for these materials, the filler particle network is a bulk object, but not a linear one).

TABLE 1.2   The dimensions of nanofiller particle structure (aggregates of particles) in elastomeric nanocomposites.

Nanocomposite	D_n from Eq. (1.23)	D_n from Eq. (1.25)	d_0	d_{surf}	ϕ_n	D_n from Eq. (1.29)
BSR/TC	1.19	1.17	2.86	2.64	0.48	1.11
BSR/nanoshungite	1.10	1.10	2.81	2.56	0.36	0.78
BSR/microshungite	1.36	1.39	2.41	2.39	0.32	1.47

Another method of D_n experimental determination uses the so-called 'quadrat method'. In essence this consists of the following. On the en-larged nanocomposite microphotograph (see Fig. 1.2) a net of quadrates with a quadrate side size of α_i, changing from 4.5 mm up to 24 mm with a constant ratio of $\alpha_{i+1}/\alpha_i = 1.5$, is applied and then the quadrates number N_i, in to which the nanofiller particles hit (fully or partly), is added up. Five arbitrary net positions for microphotography were chosen for each mea-surement. If the nanofiller particles network is a fractal, then Eq. (1.25) should be fulfilled:

$$N_i \sim S_i^{-D_n/2} \tag{1.25}$$

where: S_i is the quadrate area, which is equal to α_i^2.

In Fig. 1.6, the dependencies of N_i on S_i in double logarithmic co-ordi-nates for the three nanocomposites studied, corresponding to Eq. (1.25), is shown. As one can see, these dependencies are linear, which allows the determination of the value of D_n from their slope. The values of D_n determined according to Eq. (1.25) are also shown in Table 1.2, from which a good agreement of the dimensions of D_n, obtained by the two previously described methods, follows (their average discrepancy is 2.1% after these dimensions were recalculated for three-dimensional space according to Eq. (1.24)).

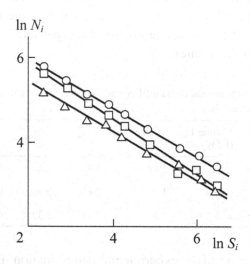

FIGURE 1.6 The dependencies of covering quadrates number N_i on their area S_i, corresponding to Eq. (1.25), in double logarithmic co-ordinates for nanocomposites on the basis of BSR. The designations are the same as those given in Fig. 1.5.

In Eq. (1.25) the condition should be fulfilled as follows:

$$N_i - N_{i-1} \sim S_i^{-D_n} \tag{1.26}$$

In Fig. 1.7 the dependence, corresponding to Eq. (1.26), for the three elastomeric nanocomposites studied is shown. As one can see, this dependence is linear, passes through the origin, which according to Eq. (1.26) is confirmed by nanofiller particle (aggregates of particles) 'chains,' which are self-similar within the selected α_i range. It is obvious, that this self-similarity will be a statistical one. It should be noted, that the points, corresponding to $\alpha_i = 16$ mm for nanocomposites: BSR/TC and BSR/microshungite, do not correspond to a straight line.

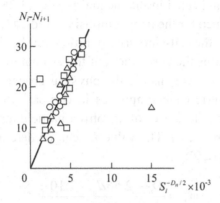

FIGURE 1.7   The dependences of $(N_i - N_{i+1})$ on the value of corresponding to Eq. (1.26), for nanocomposites based on BSR. The designations are the same, as those in Fig. 1.5.

The electron microphotographs in Fig. 1.2 give the self-similarity range for nanofiller 'chains' of 464–1472 nm. For the nanocomposite: BSR/nanoshungite, which has no points deviating from a straight line (see Fig. 1.7), the α_i range is 311–1510 nm, which agrees well enough with the self-similarity range indicated previously.

It should be noted that the minimum range of the measurement scales of S_i should contain at least one self-similarity iteration. In this case the condition for the ratio of the maximum S_{max} and minimum S_{min} areas of covering quadrates should be fulfilled by this equation:

$$\frac{S_{max}}{S_{min}} > 2^{2/D_n}$$

(1.27)

Thus, accounting for the restriction defined previously, $S_{max}/S_{min} = 121/20.25 = 5.975$, which is larger than values of for the nanocomposites studied, which are equal to 2.71–3.52. This means that the measurement scales is chosen correctly.

The self-similarity iterations number m, can be estimated from the inequality:

$$\left(\frac{S_{max}}{S_{min}}\right)^{D_n/2} > 2^{\mu}$$

(1.28)

Using the values indicated in the inequality (Eq. (1.28)) parameters, m = 1.42–1.75 is obtained for the nanocomposites studied (i.e., in these experimental conditions, the self-similarity iteration number is larger than unity, which again confirms the correctness of the estimation of the value D_n).

In conclusion, let us consider the physical grounds of smaller values of D_n for elastomeric nanocomposites in comparison with polymer microcomposites (i.e., the causes of nanofiller particle formation in the elastomeric nanocomposites). The value D_n can be determined theoretically according to Eq. (1.29):

$$\varphi_{if} = \frac{D_n + 2.55d_0 - 7.10}{4.18} \tag{1.29}$$

where: φ_{if} is the interfacial region's relative fraction, and d_0 is nanofiller initial particle's surface dimension.

The dimension d_0 estimation can be carried out with the help of Eq. (1.4) and the value φ_{if} can be calculated according to Eq. (1.7). The results of the dimension D_n theoretical calculation according to Eq. (1.29) are shown in Table 1.2, from which the theoretical and experimental results show good agreement. Eq. (1.29) indicates unequivocally the cause of a filler's different behavior in nano and microcomposites. The high (close to 3, see Table 1.2) values of d_0 for nanoparticles and relatively small ($d_0 = 2.17$ for graphite) values of d_0 for microparticles at comparable values of φ_f will be more discussed in the following sections.

Thus, the results stated previously have shown, that nanofiller particle (aggregates of particles) 'chains' in elastomeric nanocomposites are physically fractal within the self-similarity (and, thus, fractality) range of ~500–1450 nm. In this range their dimension D_n can be estimated according to Eqs. (1.23), (1.25) and (1.29). The cited examples demonstrate the necessity of choosing the correct range of measurement scales. As it has been noted earlier, the linearity of the plots, corresponding to Eqs. (1.23) and (1.25), and the D_n nonintegral value do not guarantee object self-similarity (and, thus, fractality). The low dimensions of the nanofiller particle (aggregates of particles) structure are due to the high fractal dimension of the initial nanofiller particle's surface.

The histogram in Fig. 1.8 shows elastic modulus (E) change, obtained in nano-indentation tests, as a function of load on indenter P or nano-indentation depth (h). Since for all three of the nanocomposites considered the dependencies $E(P)$ or $E(h)$ are identical qualitatively, then the dependence $E(h)$ was chosen for nanocomposite, which reflects the indicated scale effect quantitative aspect in the clearest way.

In Fig. 1.9, the dependence of E on h_{pl} (see Fig. 1.10) is shown, which breaks down into two linear parts. Such dependencies for elastic modulus – strains are typical for polymer materials in general and are due to intermolecular bonds anharmonicity. It has also been shown that the dependence of $E(h_{pl})$ – the first part at $h_{pl} \leq 500$ nm is not connected with the relaxation processes and has a purely elastic origin. The elastic modulus (E) on this part changes in proportion to h_{pl} as shown by this equation:

$$E = E_0 + B_0 h_{pl} \tag{1.30}$$

where: E_0 is 'initial' modulus, that is, modulus, extrapolated to $h_{pl} = 0$, and the coefficient B_0 is a combination of the first and second elastic constants (for the case $B_0 < 0$).

Further the Grüneisen parameter (g_L), characterizing the intermolecular bonds anharmonicity level, can be determined as follows:

$$\gamma_L \approx -\frac{1}{6} - \frac{1}{2}\frac{B_0}{E_0}\frac{1}{(1-2\nu)} \tag{1.31}$$

where: n is the Poisson ratio, accepted for elastomeric materials as being equal to ~0.475.

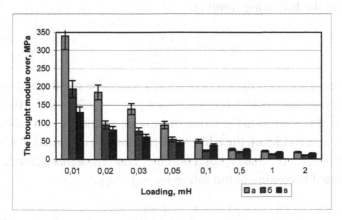

FIGURE 1.8 The dependencies of reduced elastic modulus on load on indentor for nanocomposites using BSR, filled with (a) TC (b) micro and (c) nanoshungite.

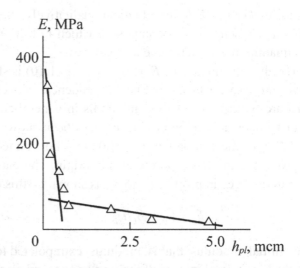

FIGURE 1.9 The dependence of reduced elasticity modulus (E), obtained in a nano-indentation experiment, on plastic strain (h_{pl}) for nanocomposites BSR/TC.

Calculation according to Eq. (1.31) has given the following values: g_L is 13.6 for the first part and 1.50 for the second one. Let us note that the first of the g_L values shown is typical for intermolecular bonds, whereas the second value of g_L is much closer to the corresponding value of the Grüneisen parameter (G) for intrachain modes.

Poisson's ratio (n) can be estimated by known values for g_L (or G) according to the following formula:

$$\gamma_L = 0.7\left(\frac{1+v}{1-2v}\right) \tag{1.32}$$

The estimations according to Eq. (1.32) gave: for the dependence $E(h_{pl})$ for the first part n = 0.462, for the second part n = 0.216. If for the first part the value of n is close to the value of Poisson's ratio for non-filled rubber, then for the second part, the additional estimation is required. As is known, the Poisson's ratio value (v_n) of a polymer composite (nanocomposite) can be estimated according to the following equation:

$$\frac{1}{v_n} = \frac{\varphi_n}{v_{TC}} + \frac{1-\varphi_n}{v_m} \tag{1.33}$$

where: φ_n is nanofiller volume fraction, v_{TC} is the Poisson's ratio of the nanofiller (TC), and v_m is the Poisson's ratio of the polymer matrix.

The value v_m is accepted to be equal to 0.475 and the magnitude v_{TC} is estimated as follows: the nanoparticle TC aggregates fractal dimension (d_f^{ag}) value is equal to 2.40 and then the value of v_{TC} can be determined according to Eq. (1.34):

$$d_f^{ag} = (d-1)(1+v_{TC}) \qquad (1.34)$$

Using Eq. (1.34) $v_{TC} = 0.20$ and calculation of v_n using Eq. (1.33) gives the value of 0.283, which is close enough to the value of n = 0.216 according to the estimation of Eq. (1.32). A comparison of the values of n and v_n obtained by the indicated methods demonstrates, that in the dependence $E(h_{pl})$ (h_{pl} <0.5 mcm), the first part of the nano-indentation tests, only the rubber-like polymer matrix (n = v_m ≈0.475) is included and in the second part, taking the entire nanocomposite as a homogeneous system: n = v_n ≈0.22.

Let us consider further E reduction at h_{pl} growth (Fig. 1.9) within the framework of density fluctuation theory, of which the value y can be estimated as follows:

$$\psi = \frac{\rho_n kT}{K_T} \qquad (1.35)$$

where: ρ_n is nanocomposite density, k is Boltzmann constant, T is testing temperature, K_T is the isothermal modulus of dilatation, connected with Young's modulus (E) by the following equation:

$$K_T = \frac{E}{3(1-v)} \qquad (1.36)$$

In Fig. 1.10, the calculation of the volume of the material deformed at nano-indentation (V_{def}) using the Berkovich indentor is shown and in Fig. 1.11 the dependence y(V_{def}) using logarithmic coordinates is shown. As it follows from the data of this Fig., the density fluctuation growth is observed as the value of V_{def} increases. Extrapolating the plot of y(ln V_{def}) to y = 0 gives ln V_{def} ≈13 or $V_{def}(V_{def}^{cr})$ = 4.42 × 10⁵ nm³. Having determined the linear scale l_{cr} of transition to y = 0 as $(V_{def}^{cr})^{1/3}$, we obtain l_{cr} = 75.9 nm, which is close to the upper boundary of the nanosystem's dimensional range which is equal to 100 nm. Thus, the results stated previously, suppose that nanosystems are systems, in which density fluctuations are absent, whereas they are always taking place in microsystems.

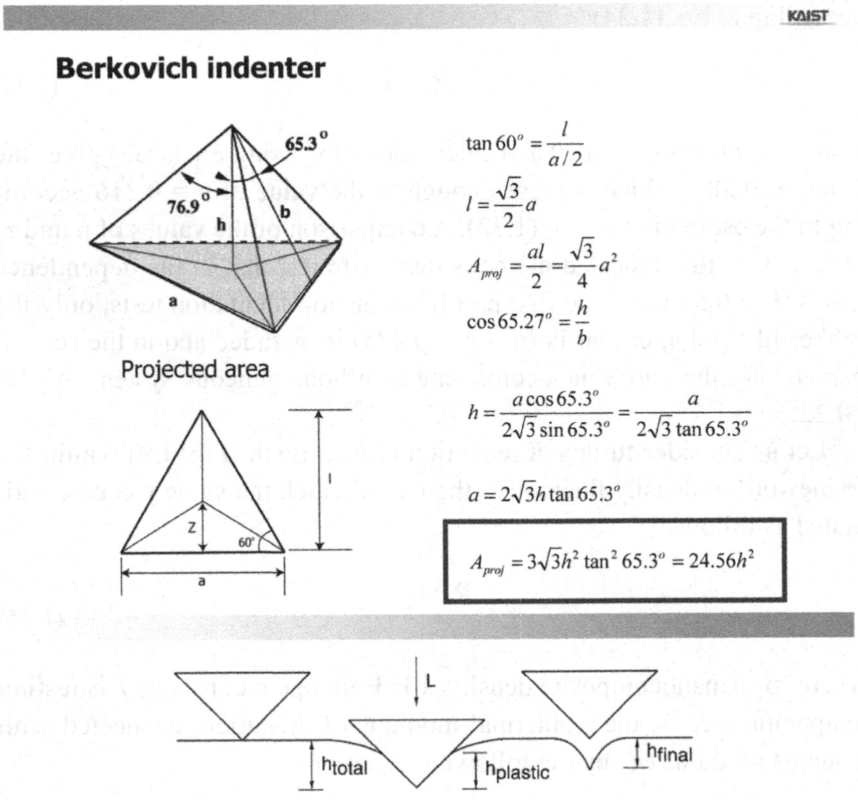

Berkovich indenter

$$\tan 60^{\circ} = \frac{l}{a/2}$$

$$l = \frac{\sqrt{3}}{2} a$$

$$A_{proj} = \frac{al}{2} = \frac{\sqrt{3}}{4} a^2$$

$$\cos 65.27^{\circ} = \frac{h}{b}$$

$$h = \frac{a\cos 65.3^{\circ}}{2\sqrt{3}\sin 65.3^{\circ}} = \frac{a}{2\sqrt{3}\tan 65.3^{\circ}}$$

$$a = 2\sqrt{3}h\tan 65.3^{\circ}$$

$$A_{proj} = 3\sqrt{3}h^2 \tan^2 65.3^{\circ} = 24.56h^2$$

FIGURE 1.10 The schematic image of a Berkovich indentor and nano-indentation process.

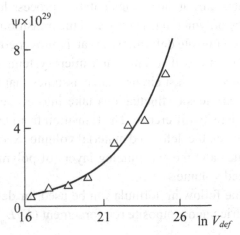

FIGURE 1.11 The dependence of density fluctuation (y) on volume of deformed nanoparticles in the nano-indentation process material (V_{def}) in logarithmic coordinates for nanocomposites – BSR/TC.

It follows from the data of Fig. 1.9, that the transition from nano- to microsystems occurs within the range h_{pl} = 408–726 nm. Both the previously indicated values of h_{pl} and the corresponding values of $(V_{def})^{1/3}$ = »814–1440 nm can be chosen as the linear length scale (l_n), corresponding to this transition. From the comparison of these values of l_n with the distance between nanofiller particles aggregates (L_n) (L_n= 219.2–788.3 nm for the considered nanocomposites, see Fig. 1.3) it follows, that for transition from nano- to microsystems, l_n should include at least two nanofiller particle aggregates and surrounding them layers of polymer matrix, that is, the lowest linear scale of nanocomposite simulation as a homogeneous system. It is easy to see that the nanocomposite structure homogeneity condition is harder to obtain from the criterion y = 0.

It is obvious, that Eq. (1.35) is inapplicable to nanosystems, since $\psi \to 0$ assumes that $K_T \to \infty$, which is physically incorrect. Therefore, the value E_0, obtained by the dependence $E(h_{pl})$ extrapolation (see Fig. 1.9) to h_{pl} = 0, should be accepted as E for the nanosystems.

Thus, the results stated above have shown, that the elastic modulus change at nanoindentation for particulate-filled elastomeric nanocomposites is due to a number of causes, which can be elucidated within the frameworks of anharmonicity conception and density fluctuation theory.

In nanocomposites during the nanoindentation process local strain is realized, affecting the polymer matrix only, and the transition to macrosystems means that nanocomposite deformation is an homogeneous system. Meanwhile it should be noted that nano- and microsystems differ by density fluctuation absence. The last circumstance assumes that for the nanocomposites considered, density fluctuations take into account nanofiller and polymer matrix density difference. The transition from nano- to microsystem is realized, when the deformed material volume exceeds the nanofiller particle aggregates and the surrounding layers of polymer matrix, which give the combined volume.

Meanwhile the following formula can be used for determining the degree of elastomeric nanocomposite reinforcement (E_n/E_m):

$$\frac{E_n}{E_m} = 15.2\left[1 - \left(d - d_{surf}\right)^{1/t}\right]$$

(1.37)

where: t is the index percolation, which in this case is equal to 1.7.

From Eq. (1.37) it follows, that the nanofiller particles' (aggregates of particles) surface dimension (d_{surf}) is the parameter, controlling the degree of reinforcement of the nanocomposites. From Eqs. (1.4)–(1.6) it follows unequivocally, that the value of d_{surf} is defined only by the size of the nanofiller particles (aggregates of particles) (R_p). In turn, it follows from Eq. (1.37), that the degree of reinforcement of the elastomeric nanocomposites (E_n/E_m) is defined by the dimension d_{surf} only, or by the size R_p only. This means, that the reinforcement effect is controlled by the nanofiller particle (aggregates of particles) size only and this is the true nano-effect.

In Fig. 1.12, the dependence of E_n/E_m is shown, which corresponds to Eq. (1.37), for nanocomposites with different elastomeric matrices (natural rubber (NR) and BSR, accordingly) and different nanofillers. Despite the indicated differences in composition, all the data shown are described well by Eq. (1.37).

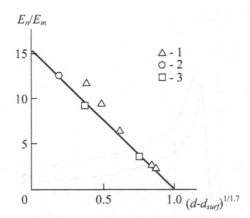

FIGURE 1.12   The dependence of degree of reinforcement (E_n/E_m) on the value of the parameter $(d-d_{surf})^{1/1.7}$ for nanocomposites: (1) NR/TC (2) BSR/TC and (3) BSR/shungite.

In Fig. 1.13, two theoretical dependences of E_n/E_m on nanofiller particle size (diameter D_p), calculated according to Eqs. (1.4)–(1.)6 and (1.37), are shown. However, for the curve 1 calculation, the value of D_p for the initial nanofiller particles was used and for curve 2 calculation – the nanofiller particles' aggregate size (see Fig. 1.3). As was expected, the growth of E_n/E_m at D_p or decreased in addition the calculation using D_p (nonaggregated nano-filler) gives higher E_n/E_m values in comparison with using the aggregated one (D_p^{ag}). At D_p £50 nm faster growth of the E_n/E_m at D_p reduction is observed than at $D_p > 50$ nm, this was also expected. In Fig. 1.13, the critical theoretical value for this transition, calculated according to the previously indicated general principles, is indicated by a vertical dotted line. In conformity with these principles the nanoparticles' size in a nanocomposite is determined according to the condition, when the division surface fraction in the entire nanomaterial volume makes up about 50% or more. This fraction is estimated approximately by the ratio $3l_{if}/D_p$, where l_{if} is interfacial layer thickness. As was noted previously, the data from Fig. 1.1 gave the average experimental value of $l_{if} = $ »8.7 nm. Furthermore from the condition $3l_{if}/D_p = $ ≈0.5 we obtain $D_p = $ »52 nm, that is shown in Fig. 1.13 by a vertical dotted line. As was expected, the value $D_p = $ »52 nm is for regions of slow ($D_p > 52$ nm) and fast ($D_p \leq 52$ nm) E_n/E_m growth at D_p decreased . In other words, the materials with nanofiller particles of size $D_p = \leq 52$ nm should be considered as true nanocomposites.

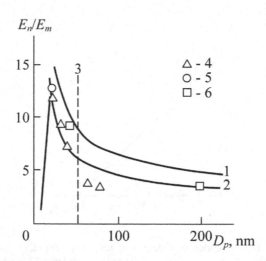

FIGURE 1.13 The theoretical dependences of degree of reinforcement (E_n/E_m) on nanofiller particles of size D_p, calculated according to Eqs. (1.4)–(1.6) and (1.37) (1) with initial nanoparticles (2) nanoparticle aggregates using size (3) the boundary value (D_p), corresponding to the true nanocomposite, the experimental data for nanocomposites (4) NR/TC (5) BSR/TC and (6) BSR/shungite.

Let us note in conclusion, that although the curves 1 and 2 of Fig. 1.13 are similar, nanofiller particles aggregation, which curve 2 represents, decreases as the degree of reinforcement increases. At the same time the experimental data corresponds exactly to curve 2, which was to expected because of the aggregation processes, which always take place in real composites. The values of d_{surf} obtained according to Eqs. (1.4)–(1.6), correspond well to the ones determined experimentally. So, for nanoshungite and two types of TC, the calculation by the method indicated gives the following d_{surf} values: 2.81, 2.78 and 2.73, whereas the experimental values of this parameter are equal to: 2.81, 2.77 and 2.73, respectively, that is, practically full agreement of theoretical and experimental results was obtained.

13.2.3 CONCLUDING REMARKS

The aggregation of the initial nanofiller powder particles in more or less large particle aggregates always occurs in the course of the process of making particulate-filled polymer composites in general and elastomeric nanocomposites in particular. The aggregation process depends on composites (nanocomposites) macroscopic properties. For nanocomposites, the nanofiller aggregation process gains special significance, as nanofiller particles aggregates size exceeds 100 nm. At present several methods exist, which allow the suppression of the nanoparticle aggregation process. Proceeding from this, in the present chapter, the theoretical treatment of dispersion of the nanofiller aggregation process in a BSR matrix within the frameworks of irreversible aggregation models was carried out.

A nanofiller disperse particle aggregation process in elastomeric matrix was studied. The modified model of irreversible aggregation particle-clusters was used for this process of theoretical analysis. The modification necessary is defined by the simultaneous formation of a large number of nanoparticle aggregates. The approach offered here, allows prediction of nanoparticle aggregates' final parameters as a function of the initial particle size, their contents and a number of other factors.

The elastomeric particulate-filled nanocomposite based on BSR was used in the study. Mineral shungite, nanodimensional and microdimensional particles and also industrially produced TC with mass contents of 37 mass% were used as a filler. The analysis of the shungite particles received from the milling process were monitored with the aid of an analytical disk centrifuge, allowing the determination of the size and distribution by sizes within certain ranges and with high precision.

- The nanostructure was studied on atomic force microscopy and by semicontact method in the force modulation regime. Atomic force microscopy results were processed with the aid of a specialized software package SPIP. This software processes the whole number functions, which are necessary for precise images analysis. Thus, the results, discussed previously have shown, that the elastomeric reinforcement effect is the true nano-effect, which is defined by the initial nanofiller particle size only. The indicated particle aggregation, which is always taking place in real materials, only changes the degree of reinforcement quantitatively, by reducing it. For the nanocomposites considered, the upper size limit of nanoparticles is ~ 52 nm.

The experimental analysis of particulate-filled nanocomposites of BSR/fullerene-containing mineral (nanoshungite) was fulfilled with the aid of atomic force microscopy, nano-indentation methods and computer analysis of the results. The theoretical analysis was carried out within the frameworks of fractal analysis. It has been shown that the interfacial regions in the nanocomposites used are actually the same reinforcing elements as nanofillers. The conditions of the transition from nano- to microsystems were discussed. The fractal analysis of nanoshungite particles aggregation in the polymer matrix was performed. It has been shown that reinforcement of the studied nanocomposites is a true nano-effect.

Attempts have been made to improve the mechanical properties of polymer-based materials, by adding a percentage of selected filler particles. There has been considerable improvement of properties such as elastic modulus, fracture toughness, flexural strength and hardness with the increase of the filler volume.

13.3 CASE STUDY II—VISCOMETRIC FLOW FOR PARTICULATE FILLED NANOPOLYMERS

Improving material properties and creating more specific tailored properties have become more important over the last decades. Combining different materials to benefit from their usually very different properties creates better materials with the combined properties. Composite materials have been around for ages and have already proved their use.

In recent years an increasing interest has been shown in nanocomposites. By choosing fillers with at least one dimension in the nanometer range, the surface to volume ratio increases tremendously. Thus, the interface between filler and matrix material increases, creating a bigger impact of the filler properties on the overall properties, such as higher stiffness, higher melt strength, lower permittivity, and improved barrier properties.

Due to the positive influence of these nanofillers in the nanocomposites, an abundance of articles on different methods to quantify the influence of the nanofiller can be found in the literature. Many articles on assessing the clay dispersion in a polymer matrix by morphological and rheological studies have been published. Due to the relatively easy sample preparation and sample loading, rheology is often used to screen or characterize the nanofiller dispersion, or more generally determine the influence of the nanofiller on the overall rheological behavior of (thermoplastic) nanocomposites.

This brings in the need to study the behavior of polymeric liquid in simple flows and for simple systems, with the hope that the knowledge gained can be appropriately used in a complex flow pattern. The word rheology is defined as the science of deformation and flow. Rheology involves measurements in controlled flow, mainly the viscometric flow in which the velocity gradients are nearly uniform in space. In these simple flows, there is an applied force where the velocity (or the equivalent shear rate) is measured or *vice versa*.

Filled polymers exhibit a diverse range of rheological properties, varying from simple viscous fluids to highly elastic solids with increasing filler volume fraction. The effect of filling on rheology is well-known for the small volume fraction range, where the reinforcement could be attributed to hydrodynamic effects caused by the solid inclusions in the melt stream. For a high volume fraction where direct particle contacts dominate the deformation, a straightforward solution of hydrodynamic equations is difficult and theoretical models based on realistic structural ideas are so far missing [1–8].

However, filled polymers usually show strong flow as well as strain and temperature history dependent rheological behaviors. It is always important to determine the dynamic viscoelastic properties at a strain that are low enough not to affect the material response.

Small strain-amplitude frequency sweep is usually used to collect linear rheological data, which are reproduced for repeated measurements within a certain experimental error. Generally, rheology is used to assess the state of dispersion of fillers in the melt.

In highly filled polymers, solid-like yielding can be observed even at temperatures above the quiescent melting temperature (T_m) or glass transition temperature (T_g) of the polymer [9–13].

It should be noted that dynamic rheology in the linearity regime is sensitive to filler dispersion in polymers. However, a straightforward description of how linear rheology varies with volume fraction is still missing so far [14–20].

Inorganic nanofiller of various types' usage for polymer nanocomposites production have been widely spread. However, the nanomaterials melt properties already mentioned are not studied completely enough. As a rule, when the application of nanofillers is considered, then a compromise between mechanical properties in solid phase melt viscosity at processing enhancement, nanofillers dispersion problem and process economic characteristics is achieved. Proceeding from this, the relationship between nanofiller concentration and geometry and nanocomposites melt properties is an important aspect of polymer nanocomposites study. Therefore, the purpose of this chapter is an investigation and theoretical description of the dependence of nanocomposite, high-density polyethylene (HDPE)-calcium carbonate ($CaCO_3$) melt viscosity on nanofiller concentration [20–31].

13.3.1 EXPERIMENTAL

HDPE of industrial production was used as a matrix polymer and nano-dimensional $CaCO_3$ in the form of compound with a particle size of 80 nm and mass content 1–10 mass% was used as a nanofiller.

HDPE-$CaCO_3$ nanocomposites were prepared by mixing (melting) the individual components in a twin-screw extruder. Mixing was performed at a temperature of 210–220 °C with a screw speed of 15–25 rpm over 5 min. Test samples were obtained using a casting under pressure method on casting machine at 200°C and a pressure of 8 MPa.

The nanocomposite viscosity was characterized by the melt flow index (MFI). The measurements of MFI were performed on an extrusion-type plastometer with a capillary diameter of 2.095 ± 0.005 mm, at a temperature of 240°C and a load of 2.16 kg. This sample was maintained at the indicated temperature for 4.5 ± 0.5 minutes.

Uniaxial tension mechanical tests were performed on the samples shaped like a two-sided spade with different sizes. Tests were conducted on a universal testing apparatus at a temperature of 20°C and a strain rate of $\sim 2 \; ' \; 10^{-3} \; s^{-1}$.

13.3.2 RESULTS AND DISCUSSION

For polymer microcomposites (i.e., composites with micron sized filler), two simple relationships between melt viscosity (h), shear modulus (G) in solid-phase state and degree of filling volume (φ_n) were obtained. The relationship between h and G has the following form:

$$\frac{\eta}{\eta_0} = \frac{G}{G_0} \qquad (2.1)$$

where: η_0 is the melt viscosity of matrix polymer, and G_0 is the shear modulus of matrix polymer.

The microcomposite melt viscosity increase can be estimated as follows (for $\varphi_n < 0.40$):

$$\frac{\eta}{\eta_0} = 1 + \phi_n \qquad (2.2)$$

In Fig. 2.1, the dependencies of the ratios of G_n/G_m and η_n/η_m, where G_n and η_n are shear modulus and melt viscosity of nanocomposite, G_m and η_m are the same characteristics for the initial matrix polymer, on $CaCO_3$ mass content (W_n) for nanocomposites HDPE-$CaCO_3$. Shear modulus G was calculated according to the following general relationship:

$$G = \frac{E}{d_f} \qquad (2.3)$$

where: E is Young's modulus, d_f is nanocomposite structure fractal dimension, determined according to the equation:

$$d_f = (d-1)(1+v) \qquad (2.4)$$

where: d is a dimension of Euclidean space, in which a fractal was considered (it is obvious, that in our case $d = 3$); s is normal stress, n is Poisson's ratio, estimated by mechanical test results by this relationship:

$$\frac{\sigma_Y}{E} = \frac{1-2v}{6(1+v)} \qquad (2.5)$$

The data shown in Fig. 2.1 clearly demonstrates that for the nanocomposites studied, Eq. (2.1) is not fulfilled either qualitatively or quantitatively: the ratio of η_n/η_m decay at W_n growth corresponds to G_n/G_m enhancement and η_n/η_m absolute values are much larger than the magnitude of the corresponding G_n/G_m.

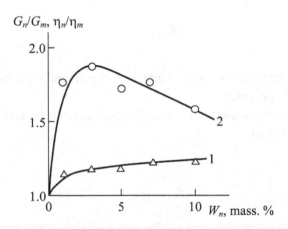

FIGURE 2.1   The dependencies of (1) shear modulus G_n/G_m and (2) melt viscosity η_n/η_m ratios of the nanocomposite G_n, η_n and matrix polymer G_m, η_m on nanofiller mass contents (W_n) for nanocomposites of HDPE-CaCO$_3$.

In Fig. 2.2, the comparison of the parameters η_n/η_m and $(1 + \varphi_n)$ for nanocomposites of HDPE-CaCO$_3$ is shown. The discrepancy between the experimental data and Eq. (2.2) is obtained again: the absolute values η_n/η_m and $(1 + \varphi_n)$ discrepancy is observed and $(1 + \varphi_n)$ enhancement corresponds to the relative melt viscosity reduction. For the graph in Fig. 2.2 the nominal φ_n value was used, which does not take into consideration the aggregation of the nanofiller particles and is estimated according to the equation:

$$\phi_n = \frac{W_n}{\rho_n} \qquad (2.6)$$

where: ρ_n is nanofiller particles density, which was determined according to the formula:

$$\rho_n = 0.188(D_p)^{1/3} \qquad (2.7)$$

where: D_p is $CaCO_3$ initial particles' diameter.

Thus, the data shown in Fig. 2.1 and Fig. 2.2 show that Eqs. (2.1) and (2.2) are well justified in case of polymer microcomposites are incorrect for nanocomposites. Let us consider this problem in more detail. Using Eq. (2.1) and Kerner's equation for G and η_n following equation is obtained:

$$\frac{\eta_n}{\eta_m} = 1 + \frac{2.5\phi_n}{1 - \phi_n} \qquad (2.8)$$

Since the value of h is inversely proportional to the MFI, then Eq. (2.8) can be rewritten as:

$$\frac{MFI_m}{MFI_n} = 1 + \frac{2.5\phi_n}{1 - \phi_n} \qquad (2.9)$$

where: MFI_m is the MFI value for the matrix polymer, and, MFI_n is the MFI value for the nanocomposite.

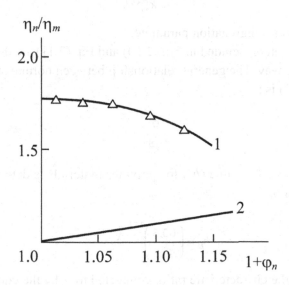

FIGURE 2.2 The dependence of nanocomposite and matrix polymer melt viscosities ratio (η_n/η_m) on nanofiller volume contents $(1 + \phi_n)$ (1) for nanocomposites HDPE-$CaCO_3$.

Three methods can be used for the estimation of φ_n in Eqs. (2.8) and (2.9). The first of these was described previously, which gives a nominal value of φ_n. The second method is usually applied for microcomposites, when a massive filler density is used as ρ_n, that is, ρ_n = constant ≈ 2000 kg/m³ for CaCO₃. And lastly, the third method also uses Eq. (2.6) and Eq. (2.7), but it takes into consideration nanofiller particle aggregation and in this case in Eq. (2.7), the initial nanofiller particle diameter (D_p) is replaced by the particles' aggregate diameter (D_{ag}). The degree of aggregation of the CaCO₃ nanoparticles and, thus, the D_{ag} value can be estimated using the dispersive strength theory, where yield stress at shear (τ_n) of nanocomposite, is determined as follows:

$$\tau_n = \tau_m + \frac{G_n b_B}{\lambda} \qquad (2.10)$$

where: τ_m is yield stress at shear of polymer matrix, b_B is Burgers' vector, and l is distance between nanofiller particles.

For the nanofiller particles' aggregation Eq. (2.10) assumes:

$$\tau_n = \tau_m + \frac{G_n b_B}{k(\rho)\lambda}, \qquad (2.11)$$

where: $k(r)$ is an aggregation parameter.

The parameters included in Eq. (2.10) and Eq. (2.11) are determined in the following way. The general relationship between normal stress (s) and shear stress (t) is :

$$\tau = \frac{\sigma}{\sqrt{3}} \qquad (2.12)$$

The Burger's vector value (b_B) for polymer materials is determined from the relationship:

$$b_B = \left(\frac{60.5}{C_\infty} \right)^{1/2} Å \qquad (2.13)$$

where: C_∞ is the characteristic ratio, connected to d_f by the equation:

$$C_\infty = \frac{2d_f}{d(d-1)(d-d_f)} + \frac{4}{3} \qquad (2.14)$$

And finally, the distance (l) between nonaggregated nanofiller particles is determined by the equation:

$$\lambda = \left[\left(\frac{4\pi}{3\phi_n} \right)^{1/3} - 2 \right] \frac{D_p}{2} \qquad (2.15)$$

From Eq. (2.11) and Eq. (2.15) $k(r)$ growth is from 5.5 up to 11.8 in the range of $W_n = 1-10$ mass% for the nanocomposites studied. Let us consider, how such $k(r)$ growth is reflected on the nanofiller particles' aggregates diameter (D_{ag}). Combining Eqs. (2.6), (2.7) and (2.15) gives the expression:

$$k(\rho)\lambda = \left[\left(\frac{0.251\pi D_{ag}^{1/3}}{W_n} \right)^{1/3} - 2 \right] \frac{D_{ag}}{2} \qquad (2.16)$$

Calculation according to the Eq. (2.16) shows a D_{ag} increase (corresponding to $k(r)$ growth) from 320 up to 580 nm in the indicated range of W_n. Furthermore, the real value of ρ_n for the aggregated nanofiller can be calculated according using Eq. 2.7 and the real degree of filling (ϕ_n) can be calculated using Eq. (2.6). Figure 2.3 shows a comparison of the dependencies of $MFI_n(W_n)$, obtained experimentally and calculated according to Eq. (2.9) using the values of ϕ_n, estimated by the three methods discussed previously. As one can see, the theoretical results obtained using Eq. (2.9) do not agree with the experimental data either qualitatively or quantitatively.

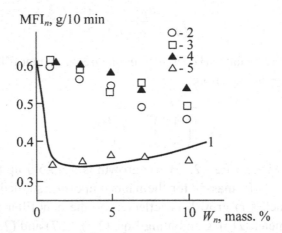

FIGURE 2.3 The dependencies of the melt flow index (MFI$_n$) on nanofiller mass contents (W_n) for nanocomposites of HDPE-CaCO$_3$. (1) – Experimental data (2–4) – data calculated from Eq. (2.9) (2) average rate (3) normal rate (4) nanofiller particles' aggregation at the condition ρ_n = constant (5) calculation according to Eq. (2.17).

The discrepancy indicated requires the application of principally differing approaches to the polymer nanocomposites' melt viscosity description. Such an approach can be fractal analysis, within the framework of which, the authors used the following relationship for fractal liquid viscosity (h) estimation:

$$\eta(l) \sim \eta_0 l^{2-d_f} \tag{2.17}$$

where: l is characteristic linear scale of flow; η_0 is constant; and d_f is fractal dimension.

Since the indicated aggregate surface comes into contact with the polymer, then its fractal dimension (d_{surf}) was chosen as d_f. The indicated dimension can be calculated as follows. The value of nanofiller particles' aggregate, specific surface (S_u) was estimated according to the equation:

$$S_u = \frac{6}{\rho_n D_{ag}} \tag{2.18}$$

and then the dimension, d_{surf}, was calculated with the aid of the equation:

$$S_u = 410 \left(\frac{D_{ag}}{2} \right)^{d_{surf}-d}.$$

(2.19)

As earlier, the value h was considered as the reciprocal value of MFI_n and the constant η_0 was accepted as being equal to $(MFI_m)^{-1}$.

Equation (2.17) allows us to make a number of conclusions. So, at the conditions mentioned previously, conservation D_{ag} increase, that is, initial nanoparticles' aggregation intensification, results to nanocomposite melt viscosity reduction, whereas d_{surf} enhancement, i.e., increasing the nanoparticles' degree of surface roughness, raises the melt viscosity. At $d_{surf} = 2.0$, i.e., the nanofiller particles have a smooth surface, the melt viscosity for the matrix polymer and the nanocomposite will be equal. It is interesting that the extrapolation of the MFI_n dependence, obtained experimentally, and for the one calculated using Eq. (2.19), d_{surf} values give the value of $MFI_n = 0.602$ g/10 min at $d_{surf} = 2.0$, that is practically equal to the experimental value of $MFI_m = 0.622$ g/10 min. The indicated factors, critical ones for nanocomposites, are not taken into consideration in continuous treatment of melt viscosity for polymer composites (Eq. (2.8)).

13.3.3   CONCLUDING REMARKS

Nanotechnology refers broadly to manipulating matter at the atomic or molecular scale and using materials and structures with nanosized dimensions, usually ranging from 1 to 100 nm. Due to their nanoscale size, nanoparticles show unique physical and chemical properties such as large surface area to volume ratios or high interfacial reactivity. Increasing nanoparticles have been demonstrated to exhibit specific interactions with contaminants in water, gases, and even soils, and such properties give hope for exciting novel and improved environmental technology.

However, the small particle size also brings issues involving mass transport and excessive pressure drops when applied in a fixed bed or any other flow-through systems, as well as certain difficulties in separation and reuse, and even possible risk to ecosystems and human health caused by the potential release of nanoparticles into the environment. An effective approach to overcoming the technical bottlenecks is to fabricate hybrid nanocomposites by impregnating or coating the fine particles onto solid particles of larger size. The widely used host materials for nanocomposite fabrication include carbonaceous materials such as granular activated carbon, silica, cellulose, sands, and polymers, and polymeric hosts are a particularly attractive option, partly because of their controllable pore size and surface chemistry as well as their excellent mechanical strength for long-term use. The resultant polymer-based nanocomposite (PNC) retains the inherent properties of nanoparticles, while the polymer support materials provide higher stability, processability and some interesting improvements caused by the nanoparticle-matrix interaction. The nanoparticles generally used include: zero-valent metals, metallic oxides, biopolymers, and single-enzyme nanoparticles (SEN). These nanoparticles could be loaded onto porous resins, cellulose or carboxymethyl cellulose, chitosan, and so on. The choice of the polymeric supports is usually guided by their mechanical and thermal behavior. Other properties such as hydrophobic/hydrophilic balance, chemical stability, bio-compatibility, optical and/or electronic properties and chemical functionalities (i.e., solvation, wettability, templating effect, and so on) have to be considered when selecting the organic hosts.

The results observed in this chapter emphasize that the microcomposites rheology description models do not give adequate treatment of melt viscosity for particulate-filled nanocomposites. The correct description of the nanocomposites rheological properties can be obtained within the frameworks of viscous liquid flow fractal models. It is significant, that such an approach differs principally from the used ones to describe microcomposites. So, nanofiller particles' aggregation reduces both melt viscosity and elastic modulus of nanocomposites in the solid-phase state. For microcomposites, melt viscosity enhancement is accompanied by elastic modulus increase.

13.4   CASE STUDY III—APPLICATION OF SYNTHETIC OR NATURAL INORGANIC FILLERS

Conventionally, micro and nano-composites have been filled with synthetic or natural inorganic particles in order to improve the material properties or simply to reduce the costs. The addition of these particles sometimes imparts disadvantages to the resultant composites such as weight increase, brittleness and opacity. Nanocomposites are a new type of composite that are particle-filled polymers for which at least one dimension of the dispersed particles is in the nanometer range. In recent years polymer/layered silicate nanocomposites have been of particular interest, both in industry and in academia, as they often reveal notable enhancements in the material's properties at very low loading levels compared to virgin polymer or conventional composites [1–14].

In order to obtain good interfacial adhesion and mechanical properties, the hydrophilic clay needs to be modified prior to its introduction in most polymer matrices, which are organophilic. When nanometric dispersion of primary clay platelets is obtained, the aspect ratio of the filler particle is increased and the reinforcement effect is improved [15–25].

Polypropylene (PP) is one of the most interesting thermoplastic materials because of its low price and balanced properties. PP is the lightest polymer and has wide range applications such as packaging, fiber, automotive industry, nondurable goods and construction materials. However, due to the low polarity of PP, obtaining the exfoliated and homogenous dispersion of the silicate layer at the nanometer level in the polymer matrix is difficult. This is principally due to the fact that the silicate clays layers have polar hydroxyl groups and are compatible only with polymers containing polar functional groups [26–38].

Due to the nonpolar nature of PP, a third component known as a compatibiliser such as maleic anhydride-*grafted*-polypropylene (MA-*g*-PP) copolymer are introduced to compensate for the difference of polarity between the polymer matrix and the clay particles.

The exfoliation and dispersion of nanoclays in polyolefin matrices prepared by melt compounding depend on the organic modifier of the nanoclay, the initial interlayer spacing, the concentration of functional groups in the compatibiliser and its overall concentration in the composite, the viscosity of the plastic resin, and the operational conditions, such as screw configurations of extruders, speed (rpm), temperature, residence time, and so on [39–50].

Since Nylon-6/clay nanocomposites with excellent thermal and mechanical properties were reported by many scientists and polymer/clay nanocomposites have attracted much attention. The improvements in thermal, mechanical and flammability properties of polymer/clay nanocomposites are significantly higher than those achieved in traditional filled polymers. To date, polymer/clay nanocomposites have been successful in several kinds of polar polymers. However, for polymers with low polarity, such as polyolefins, the current results are not satisfactory due to the low compatibility between the clay and the polyolefins. PP is a highly hydrophobic and nonpolar polymer and cannot be used as a matrix for the dispersion of inorganic hydrophilic clays in order to prepare nanocomposites without a compatibilizing modification. As a result, a complete exfoliation in PP nanocomposites has not yet been obtained, and moreover, the decrease of tensile strength of PP nanocomposites still cannot be satisfactorily avoided [51–61].

The aim of this chapter is to evaluate the influence of compatibiliser and processing conditions on the dispersion behavior of nanoclays as filler in a fiber-formed PP matrix as well as the tensile strength of PP nanocomposites.

13.4.1 EXPERIMENTAL

PP homopolymer, Moplen HP552R, having a melt flow rate of 25 g/10 min, was used as-received. A polypropylene-*graft*-maleic anhydride (PP-*g*-MA) with 0.1% of MA (PRIEX 20070), with a melt flow rate of 64 g/10 min) was used as a compatibilizing agent.

Preparation of the nanocomposite samples was done using a twin-screw extruder and an internal mixer. The temperature of the twin-screw extruder was maintained at 160, 170, 180, 180, 190 and 190°C from hopper to die, and the screw speed was about 200 rpm. The reaction of an internal mixer was carried out at 180°C and 100 rpm for 15 min. Table 3.1 shows the specification of the samples prepared.

TABLE 3.1   Specification of prepared samples.

Sample	Clay (wt%)	PP-*g*-MA (wt%)	Method of preparation	Number of extrusions
PP	0	0	Direct melt mixing in an internal mixer	-
PPC	1	0	Direct melt mixing in an internal mixer	-
PPCG	1	3	Direct melt mixing in an internal mixer	-
PP1	0	0	Direct melt mixing in a twin-screw extruder	1
PP2	0	0	Direct melt mixing in a twin-screw extruder	2
PPS1–3	1	3	Direct melt mixing in a twin-screw extruder	1
PPS2–3	1	3	Direct melt mixing in a twin-screw extruder	2
PPS1–9	1	9	Direct melt mixing in a twin-screw extruder	1
PPS2–9	1	9	Direct melt mixing in a twin-screw extruder	2

The structure of layered silicates and the morphology of extruded composites specimen were analyzed by wide-angle X-ray diffraction scattering (WAXD) and scanning electron microscopy (SEM). X-ray diffraction (XRD) data were performed with a diffractometer. The X-ray beam was CuKa radiation of wavelength 1 °A with a nickel-filter operated at 40 kV and 30 mA. Diffraction spectra were obtained over a 2θ range of 2–10° at a scanning rate of 0.02°/s and the interlayer spacing (d_{001}) was calculated using Bragg's equation, $\lambda = 2d\sin\theta$, where λ is the wavelength. The sample for WAXD measurement was prepared in film by compression molding at 190°C and 1.5 MPa. The mechanical properties were evaluated by tensile modulus, tensile strength and maximum percentage strain measurements.

13.4.2   RESULTS AND DISCUSSION

XRD is a rather simple and widely used technique for the characterization of a layered-clay dispersion in a polymer matrix. WAXS analysis was used to quantify the height between adjacent silicate platelets, and subsequently to prove the widening of this distance as the matrix polymer intercalates between the galleries. Changes in the value of 2θ reflect changes in the gallery distance of the clay and allow the intercalated silicate fractions to be distinguished. Therefore, Fig. 3.1 shows the diffraction pattern for pristine OMMT, PPC and PPCG composites prepared in an internal mixer and PPS1–3, PPS2–3, PPS1–9 and PPS2–9 composites were prepared in a twin-screw extruder with a constant clay content, that is, 1 wt%. As expected, the layered-structure of nanoclay is evident. The resulting inter-layer distances are shown in Table 3.1. The pristine OMMT exhibits a sharp peak (d_{001}) at around 2.48°, corresponding to an interlayer d-spacing of 3.56 nm. The very broad feature in the 4–8° 2θ region possibly indicates ion-exchange, possibly due to high molecular weight of the surfactant. However, after melt mixing and extrusion molding, the characteristic peak is substantially weakened, denoting that delaminating or disordering of the clay may have occurred. Thus, the XRD traces suggest that the system contains both an immiscible component and a nano-dispersed component. Introducing the compatibilizing agent to the composites decreases the d_{001} diffractions of the clay. However, the characteristic peak is not weakened by increasing the ratio of compatiblizing agent to clay.

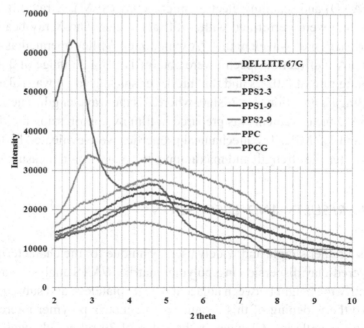

FIGURE 3.1   XRD patterns.

TABLE 3.2   XRD results and composite samples.

Sample	2θ (°)	d value (nm)
PP	2.48	3.56
PPC	2.93	3.01
PPCG	2.78	3.17

It is clearly observed that the (d_{001}) base peak position shifts into higher angles passing from 2.48° for the pristine OMMT to 2.93° and 2.78° for the PPC (Table 3.2) and PPCG composites, which indicates that PP does not intercalate into OMMT, even for the composite sample containing MA-g-PP and the broad feature in the 4–8° 2θ regions remains. It means that the clay internal spacing has decreased. In these two samples, no dispersion of clay particles occurs. It suggests that the nanocomposite structure is not achieved. This is also confirmed by SEM images showing the agglomeration of clay particles in the surface of the PPC and PPCG composites, as shown in Fig. 3.2. It is also seen that there is no significant effect of compatibiliser on the dispersion of clay particles in the polymer matrix. However, the size of agglomerated particles decreases with the introduction of the compatibilizing agent.

For the XRD patterns of PPS1–3 and PPS2–3 composites prepared in a twin-screw extruder, the broad feature in the 4–8° 2θ regions broadens with low intensity and shifts to lower angles. The d_{001} peak weakens, denoting that exfoliation or disordering of the clay may have occurred. The diffuse nature of the d_{001} peaks makes it difficult to estimate accurately their d-spacing which suggests either a formation of intercalated nanocomposites or a mixed intercalated-exfoliated system (or simply a disordered system). Thus, the XRD traces suggest that the PPS1–3 and PPS2–3 composites contain an immiscible component (from the broad feature at 4–8°) and a nano-dispersed component. Figure 3.3 shows the SEM images of agglomerated clay particles in the surface of a PPS1–3 composite, which is much less than the microcomposites prepared in an internal mixer. Figure 3.4 shows the SEM images of agglomerated clay particles in the surface of the PPS2–3 composite, which is less than in PPS1–3.

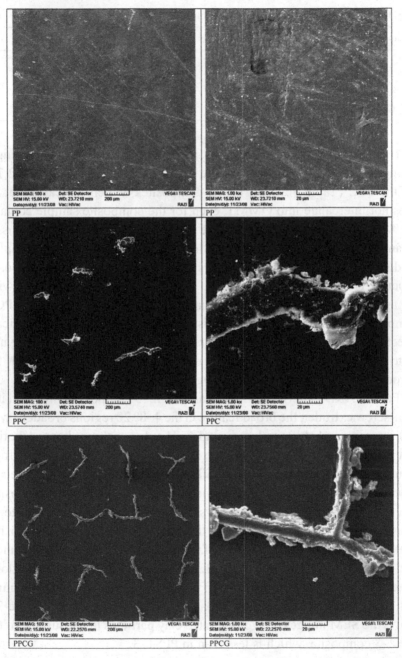

FIGURE 3.2 Surface SEM images of pristine polypropylene and composite samples prepared in an internal mixer.

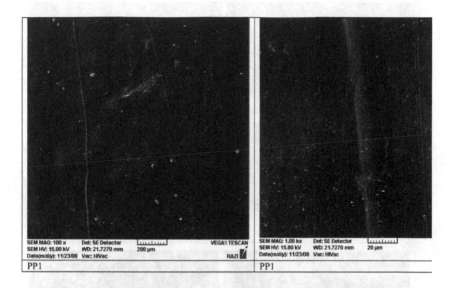

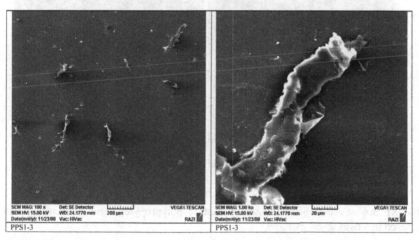

FIGURE 3.3 Surface SEM images of pristine polypropylene and composite samples prepared in a twin screw extruder.

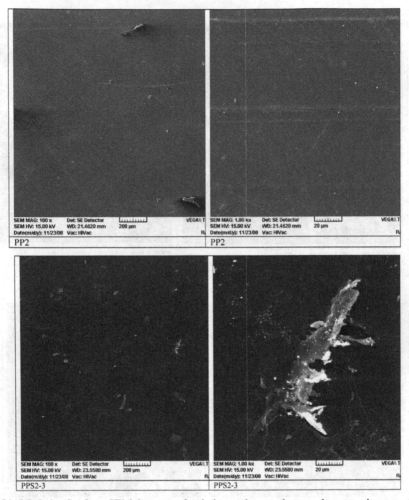

FIGURE 3.4 Surface SEM images of pristine polypropylene and composite samples prepared in a twin screw extruder.

The ability to exfoliate and disperse the layered silicate into a polymer matrix is related to a number of factors such as the exchange capacity of the silicate, the polarity of polymer and the chemical nature of the organic modifier. Processing conditions are one of the factors affecting the dispersion of the silicate layers. The disappearance of the peak in the XRD patterns and the decrease of agglomerated particles in the SEM images indicate that the dispersibility of clay particles is more pronounced in the composite samples prepared in a twin-screw extruder than in an internal mixer.

XRD investigations and SEM imaging was carried out to study the dispersion of clay in the matrix. Melt compounding in the twin-screw extruder and using the compatibilizing agent decrease the large-scale particle agglomeration. However, there is no significant change in the tensile strength of composites as shown in Table 3.3. It can be concluded that tensile strength remains the same because of the existence of an immiscible clay component. Conversely, it is worthy to note that the increase of the strength can be observed.

TABLE 3.3 Mechanical properties of the samples prepared in an internal mixer.

Sample	Modulus (MPa)	Tensile strength (MPa)	Strain (%)
PP	1850 ± 54	39.9 ± 1.8	4.3 ± 0.5
PPC	1828 ± 98	38.8 ± 0.5	4.4 ± 0.3
PPCG	1780 ± 144	37.5 ± 2.8	3.8 ± 0.1
PP1	1772 ± 159	40.6 ± 1.9	4.8 ± 0.1
PPS1–3	1690 ± 228	40.2 ± 3.5	4.8 ± 0.6
PP2	1736 ± 176	38.6 ± 2.0	4.5 ± 0.1
PPS2–3	1872 ± 52	40.3 ± 1.2	4.4 ± 0.2

13.4.3   CONCLUDING REMARKS

PP/Montmorillonite (MMT) nanoclay based composite was prepared by melt compounding with MA-g-PP in a twin-screw extruder and an internal mixer. Test specimens were prepared by compression molding. Mechanical properties such as tensile modulus, tensile strength and maximum percentage strain were measured. The clay dispersion was investigated using WAXD and SEM.

In this research the XRD patterns and SEM images show that introducing the proper compatibilizing agent ratio and suiTable processing condition like a twin-screw extruder decrease the immiscible clay component in the matrix. However, the increase of tensile strength is not achieved.

13.5   CASE STUDY IV—A MULTISCALE MICROMECHANICAL MODEL

One of the desirable end-goals of materials science research is the development of multifunctional materials. These materials are defined as compositions that bring more than one property enhancement to a particular application, thus allowing the material to replace more than one other material in an engineered object, or to replace entire classes of materials which alone, are only capable of addressing one end-use need [1–4].

The polymer nanocomposite field has been studied heavily in the past decade. However, polymer nanocomposite technology has been around for quite some time in the form of latex paints, carbon-black filled tires, and other polymer systems filled with nanoscale particles. However, the nanoscale interface nature of these materials was not truly understood and elucidated until recently [2–7]. Today, there are excellent works that cover the entire field of polymer nanocomposite research, including applications, with a wide range of nanofillers such as layered silicates (clays), carbon nanotubes/nanofibers, colloidal oxides, double-layered hydroxides, quantum dots, nanocrystalline metals, and so on. The majority of the research conducted to date has been with organically treated, layered silicates or organoclays.

A traditional composite containing micron sized or larger particles/fibers/reinforcement can best be thought of as containing two major components, the bulk polymer and the filler/reinforcement, and a third, very minor component, or interfacial polymer. Poor interfacial bonding between the bulk polymer and the filler can result in an undesirable balance of properties, or at worst, material failure under mechanical, thermal, or electrical load. In a polymer nanocomposite, since the reinforcing particle is at the nanometer scale, it is actually a minor component in terms of total weight or volume percentage in the final material. If the nanoparticle is fully dispersed in the polymer matrix, the bulk polymer also becomes a minor, and in some cases, a nonexistent part of the final material. With the nanofiller homogenously dispersed in the polymer matrix, the entire polymer becomes an interfacial polymer, and the properties of the material begin to change. Changes in properties of the interfacial polymer become magnified in the final material, and great improvements in properties are seen. Therefore, a polymer nanocomposite is a composite where filler and bulk polymer are minor components, and the interfacial polymer is the component that dictates the material's properties. With this in mind, the design of the nanoparticle is critical to the nanocomposite structure, and careful understanding of nanoparticle chemistry and structure are needed.

At present nanocomposite polymer/organoclay studies attained very big wide spreading. However, the majority of work done on this theme has mainly been of an applied character and theoretical aspects of the polymer's reinforcement by organoclays have been studied much less. In this chapter we describe a multiscale micromechanical model.

This is based on the following percolation relationship application:

$$\frac{E_n}{E_m} = 1 + 11\phi_n^{1.7} \qquad (4.1)$$

where: E_n is the elastic modulus of the nanocomposite; E_m is the elastic modulus of the matrix polymer, and ϕ_n is nanofiller volume contents.

Let us consider that the model is based on the approach, which is principally different from the micromechanical models: it is assumed that polymer composites' properties are defined by their matrix structural state only and that the role of the filler consists in modification and fixation of the matrix polymer structure.

In this chapter experimental data for the nanocomposite: epoxy polymer/Na$^+$-montmorillonite (EP/MMT) is considered.

13.5.1 EXPERIMENTAL

As matrix polymer, an epoxy polymer based on 3,4-epoxycyclohexylmethyl-3,4-epoxycyclohexane carboxylate (epoxy monomer), cured by hexahydro-4-methylphthalic anhydride at a molar ratio of 0.87–1.0 was used. To this mixture a nanofiller, MMT (Cloisite 30B) was added. Before this the nanofiller was mixed with denatured ethanol for removal of any excess of surfactants from the silicate platelets' surface. This mixture was then centrifuged at 5,000 rpm for 10 min prior to the decanting of the ethanol. After Cloisite 30B addition the mixture was stirred in a mixer at a setting of 2,500 rpm for 45 seconds and then at 3,000 rpm for the same period of time.

The nanocomposite EP/MMT samples were prepared in three stages. The samples were first cured isothermally for up to 8 h at temperatures of 80, 100, 120 and 140°C, followed by 8 h at 180°C and finally 12 h at 220°C under vacuum. The samples were cooled down at a rate of 1–2°C/ min and 20 ′ 5 ′ 1 mm^3 pieces were machined for dynamic spectroscopy measurements. MMT contents in the studied nanocomposites were 2, 5, 10 or 15 mass%.

The dynamical mechanical tests were carried out by using rheometer in torsional mode at a frequency of 1 s^{-1} in a nitrogen atmosphere.

The nanocomposite structure studies using wide-angle x-ray diffractometry were carried out on a general area diffraction detection system using copper irradiation with a wavelength (1) of 1.54 Å. The distance between MMT platelets (d_{001} interlayer spacing) was calculated according to these measurements.

13.5.2 RESULTS AND DISCUSSION

The model can be described in the following manner. Composite continuous simulation found out that property improvement depends strongly on the individual features of the filler particles: their volume fraction (f_p), particle aspect ratio (anisotropy) (L/t) and the ratio of particle and matrix mechanical properties. These important aspects of nanocomposite polymer/organoclay require coordinated and precise definition. A multiscale simulation strategy can be used for calculation of nanocomposite hierarchical morphology at a scale in the order of thousands of microns (the large particle aspect ratio within the matrix limits the structure. At a scale of several microns, the clay particle structure presents itself as either fully divided organoclay platelets with a thickness in the order of nanometers (or as organoclay parallel platelets packing, separated by interlayer galleries with a thickness of nanometers. In this case, the quantitative structural parameters, obtained by x-ray diffraction and electron microscopy methods (silicate platelets number N in the organoclay bundle, spacing d_{001} between silicate platelets) can be used for definition of the clay 'particles' geometrical features including L/t and ratio f_p to nanofiller mass content (W_n). These geometrical features together with silicate platelet stiffness estimations, can be obtained from molecular dynamics simulations, which gives a basis for the prediction of the effective mechanical properties of the organoclay particles.

Unlike many mineral fillers, used for plastics production (talcum powder, mica and so on), organoclays, in particular MMT, are capable of stratifying and dispersing into separate platelets with a thickness of about 1 nm.

The term 'intercalation' describes how small amounts of polymer penetrate into galleries between silicate platelets, which cause these platelets to separate for ~2–3 nm. Exfoliation or stratification occurs at the distance between the platelets (in x-ray techniques, this distance is called the spacing – d_{001}) is of the order of 8–10 nm. A well-stratified and dispersed nanocomposite includes separate organoclay platelets, uniformly distributed in a polymeric matrix. Equation (4.1) used for the prediction of the degree of reinforcement of nanocomposites (E_n/E_m) has shown the necessity of the following modification. First of all it has been found out that not only in the complete nanofiller (organoclay platelets), but also in the interfacial regions, formed on the platelet surface, a with relative fraction (φ_{if}) are the nanocomposite structure, reinforcing or strengthening elements. Such a situation is due to the higher stiffness of the interfacial layers in comparison with the bulk polymeric matrix by virtue of the strong interactions between polymer, organoclay and molecular mobility suppression in the layers mentioned.

Now Eq. (4.1) can be rewritten as:

$$\frac{E_n}{E_m} = 1 + 11\left(\phi_n + \phi_{if}\right)^{1.7}$$ (4.2)

Besides, within the frameworks of the fractal model of interfacial layer formation in polymer nanocomposites it has been shown that between φ_{if} and φ_n the following relationship exists:

$$\phi_{if} = 0.955\phi_n b$$ (4.3)

for intercalated organoclay and

$$\phi_{if} = 1.910\phi_n b \qquad (4.4)$$

for an exfoliated one, where b is a parameter characterizing the nanofiller – polymeric matrix interfacial adhesion level. Let us note that Eq. (4.3) and Eq. (4.4) were obtained by accounting for the strong anisotropy of the organoclay platelets. For example, for particulate (approximately spherical) nanofiller particles the similar equation has the following form:

$$\phi_{if} = 0.102\phi_n b \qquad (4.5)$$

Combining Eqs. 4.2–4.4 allows us to obtain the final variant of the formulae for nanocomposites polymer/organoclay degree of reinforcement determination:

$$\frac{E_n}{E_m} = 1 + 11\left(1.955\phi_n b\right)^{1.7} \qquad (4.6)$$

for intercalated organoclay, and

$$\frac{E_n}{E_m} = 1 + 11\left(2.910\phi_n b\right)^{1.7} \qquad (4.7)$$

for an exfoliated one.

Therefore, comparing the Eqs. (4.2)–(4.7), demonstrates that the model accounts for nanofiller volume contents, its particles' degree of anisotropy and interfacial adhesion level, but does not account for nanofiller characteristics.

The nanofiller volume fraction (ϕ_n) can be estimated as follows. At it is known, the following interconnection exists between the nanofiller volume (ϕ_n) and mass contents (W_n):

$$\phi_n = \frac{W_n}{\rho_n} \qquad (4.8)$$

where: ρ_n is the nanofiller density, determined according to the following equation:

$$\rho_n = \frac{6}{S_u D_p} \qquad (4.9)$$

where: S_u is the filler specific surface, which is equal to ~ 74 ' 10^3 m²/kg for MMT, and D_p is its particle size.

Since the MMT particle length is anisotropic and its length is ~ 100 nm, and its width is ~ 35 nm and its thickness is ~ 1 nm, then the arithmetical mean of these sizes was chosen as D_p. Then the value $\rho_n = 1790$ kg/m³.

For the nanocomposite, EP/MMT with $W_n = 2$ and 5 mass%, the interlayer spacing value $d_{001} \geq 10$ nm and for the same nanocomposites with $W_n = 10$ and 15 mass%, $d_{001} = 6.3–9.8$ nm. Therefore, according to the classification shown previously, nanocomposites with $W_n = 2$ and 5 mass% are to be defined as having exfoliated organoclay and those with $W_n = 10$ and 15 mass% are defined as intercalated organoclay. In Fig. 4.1 the curves $E_n(W_n)$, calculated for the previous cases according to Eqs. (4.6) and (4.7), at the condition $b = 1$ (perfect adhesion by Kerner) and also the experimental data for the nanocomposite, EP/MMT are shown. As one can see, there is good agreement of the data for EP/MMT with $W_n = 10$ and 15 mass% and the calculation from Eq. (4.6) and with $W_n = 2$ and 5 mass%. Therefore, above equations reflects well both the nanofiller contents and the degree of stratification of its platelets and in addition it is much simpler and physically much clearer than the multiscale model.

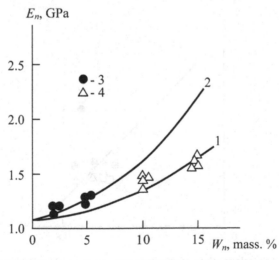

FIGURE 4.1   The comparison of calculated according to Eq. (4.6) (1) and Eq. (4.7) (2) and experimental of exfoliated (3) and intercalated (4) organoclay dependencies of elastic modulus (E_n) on organoclay mass contents (W_n) for nanocomposite EP/MMT.

One more aspect of the model, concerns the determination of parameter b, which characterizes interfacial adhesion at the nanofiller – polymeric matrix level. The condition $b = 1$ was accepted for the value of E_n (Fig. 4.1), but this parameter can also be estimated more precisely according to Eq. (4.6) and Eq. (4.7), using experimental values E_n, E_m and φ_n. An independent estimation of the interfacial adhesion level for the nanocomposites studied, can be obtained with the aid of parameter A, determined according to the equation:

$$A = \frac{1}{1-\phi_n} \frac{tg\delta_n}{tg\delta_m} - 1,$$
(4.10)

where: $tg\,\delta_n$ is the mechanical loss angle tangent for the nanocomposite, and $tg\,\delta_m$ is the mechanical loss angle tangent for the matrix polymer.

It is already known, that a decrease in A means an increase in the interfacial adhesion level. In Fig. 4.2, the comparison of parameters A and b, characterizing the interfacial adhesion level is shown for the nanocomposites EP/MMT. As has been expected, a decrease of A according to the reasons indicated previously corresponds to growth of b (determined according to Eqs. (4.6) and (4.7)) and parameter b characterizes the real level of interfacial adhesion in the nanocomposite polymer/organoclay and is close to unity.

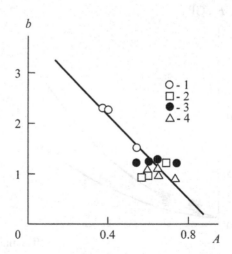

FIGURE 4.2   The relationship of parameters b and A, characterizing the interfacial adhesion level, for nanocomposite EP/MMT with an organoclay mass contents (W_n) = 2 (1), 5 (2), 10 (3) and 15 (4) mass%

Let us further discuss the organoclay 'effective particle' characteristics. We can calculate the dependencies of the silicate relative fraction (c) in an 'effective particle' and plot its dependencies on the ratio of d_{001}/d_{pl} for different N [where d_{pl} is the silicate platelet thickness ($d_{pl} = 1$ nm), and N is the platelet number in an 'effective particle']. In Fig. 4.3 the theoretical calculations are shown by solid lines for $N = 1$, 2 and 4, and experimental dependencies of $\varphi_n/(\varphi_n + \varphi_{if})$ in the EP/MMT – by points. As one can see, the similar trends of c and $\varphi_n/(\varphi_n + \varphi_{if})$ change as a function of d_{001}/d_{pl}, which indicates the identity of the 'effective particle' definition. The comparison of theoretical curves and experimental points in Fig. 4.3 shows that for the nanocomposites EP/MMT, the value of N is somewhat smaller than two (i.e., the organoclay in these nanocomposites is stratified).

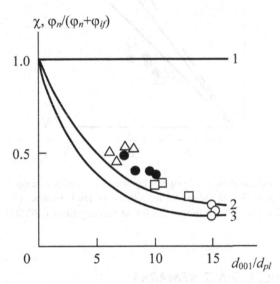

FIGURE 4.3 The theoretical dependencies of an organoclay fraction in an 'effective particle' (c) for silicate platelets number (N) = 1 (1), 2 (2) and 4 (3) and parameter $\varphi_n/(\varphi_n + \varphi_{if})$ for nanocomposite EP/MMT (points) on the value of the ratio d_{001}/d_{pl} are the same as those in Fig. 4.2.

In Fig. 4.4 the theoretical dependencies of the 'effective particles' volume fraction and the nanofiller mass content ratio (f_p/W_n) on the ratio of d_{001}/d_{pl}, are shown by straight lines for $N = 1$, 2, 3 and 4. Also in Fig. 4.4, the dependencies of the ratio of the sum $(\varphi_n + \varphi_{if})$, which in the model is analog to f_p and W_n, obtained according to Eq. (4.6) and Eq. (4.7), for the nanocomposite EP/MMT are given. The similarity of trends of the ratios as a function of d_{001}/d_{pl} change is observed again, which confirms again 'effective particle' identity. And as earlier, for the nanocomposite EP/MMT, the value of N can be estimated as being somewhat smaller than $N = 2$.

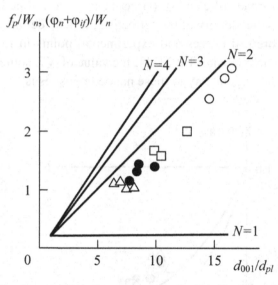

FIGURE 4.4   The dependencies of organoclay fraction in 'effective particles' volume and mass fractions ratio (f_p/W_n) in the model for silicate platelets number $(N) = 1$ (1), 2 (2), 3 (3) and 4 (4) and $(\varphi_n + \varphi_{if})/W_n$ in the model for nanocomposites EP/MMT (points) on the ratio d_{001}/d_{pl} value.

13.5.3   CONCLUDING REMARKS

Polymer clay nanocomposites are already used in many applications to enhance existing properties of a particular material, and development should be focused on the true multifunctional materials. Certainly, clay nanocomposites will continue to be used for enhanced mechanical, flammability, and gas barrier properties, but fundamental limits in clay chemistry prevent them from being used easily in applications requiring electrical/thermal conductivity or optical properties. Similarly, combinations of organoclays with other nanofillers to obtain a truly multifunctional material are will likely to be used in the future. Combining an organoclay with carbon nanotubes, or quantum dots, could yield a very interesting nanocomposite with enhanced mechanical, flammability, thermal, and electrical properties – allowing it to be a 'drop-in' replacement for many different materials in a complex part. Alternatively, the clay could enhance the properties of some existing mechanically fragile systems while keeping the other properties intact. Some early initial research has been done on combining more than one type of nanofiller in a polymer matrix, but this approach has not yet been widely studied.

Polymer nanocomposites with organoclay fillers offer improved properties and performance, providing opportunities for commercial applications. The key to significant property enhancement is to exfoliate the individual organoclay platelets into the polymer matrix to utilize their high aspect ratio and modulus. The affinity between the polymer matrix and the organoclay is one of the most important factors for determining the exfoliation level. In this chapter the model of the organoclay 'effective particle' describes the nanocomposite's polymer/organoclay structure and properties. However, the model has a clear physical significance and does not require the application of micromechanical models.

13.6   CASE STUDY V—CEMENT MATERIALS REINFORCEMENT WITH NANOPARTICLES

Composite materials reinforced with nanoparticles have recently received tremendous attention in both scientific and industrial communities due to their extraordinary enhanced properties. However, from the experimental point of view, it is a great challenge to characterize the structure and to manipulate the fabrication of nanocomposites. The development of such materials is still largely empirical and a finer degree of control of their properties has not been achieved so far. Therefore, computer modeling and simulation will play an ever-increasing role in predicting and designing material properties, and guiding such experimental work [1–18].

The use of nanoparticles in developing materials has been introduced in the last few years [1–12]. It has been observed that the extremely fine size of nanoparticles strongly affects their physical and chemical properties and enables them to be used to fabricate new materials with novelty functions. Thus, integrating nanoparticles with existing cement-based building materials may enable the fabrication of new products with some outstanding properties [2–8]. Among the nanoparticles, nanosilica (nano-SiO_2) has been used to improve the properties of cement-based materials. Some cement composites with nano-SiO_2 having excellent mechanical properties and microstructure have been reported. Scientists showed that incorporating nanoparticles into high-volume fly ash concrete can significantly increase the initial pozzolanic activity of fly ash. Scientists also concluded that nano-SiO_2 can enhance the short-term and long-term strength of high-volume, high strength concrete. It should be noted that nano-SiO_2 particles could potentially improve the negative influences caused by enhancing the compressive and flexural strength of cement concrete and mortar. As the rate of the pozzolanic reaction is proportional to the amount of surface available for reaction and because of the high specific surface of nanoparticles, they possess high pozzolanic activity. Nano-SiO_2 effectively consumes calcium hydroxide crystals (CH) which array in the interfacial transition zone between the hardened cement paste and the aggregates and produces hydrated calcium silicate (CSH) which enhances the strength of the cement paste. In addition, due to the nanoscale size of the particles, nano-SiO_2 can fill the ultra fine pores in the cement matrix. This physical effect leads to a reduction in porosity of the transition zone in the fresh concrete. This mechanism strengthens the bond between the matrix and the aggregates and improves the cement paste properties. Furthermore, it has been found that when the ultra fine particles of nano-SiO_2 uniformly disperse in the paste, they generate a large number of nucleation sites for the precipitation of the hydration products, which accelerates the cement hydration [12–32].

It has been demonstrated that rice husk ash (RHA) can be added to concrete mixtures as a substitute for the more expensive Portland cement to lower the construction costs. Compared to other agricultural by-products, RHA is high in ash and similar to silica fume – it contains considerable amounts of SiO_2. Thus, RHA is not just a cheap alternative but a well ground rice husk with most of its silica in an amorphous form and with enough specific surface so that it is very active and considerably improves the strength and durability of cement and concrete. Researchers have applied RHA and nano-SiO_2 to improve cement-based materials and have achieved great successes. However, considering using nano-SiO_2 particles in RHA mortar is an innovative approach. The main purpose of this chapter is to study the mechanical properties of RHA mortar containing nano-SiO_2 in order to further improvements of Portland cement based materials [30–52].

13.6.1 EXPERIMENTAL

In this study, ordinary Portland cement (OPC) type I, standard graded sand, RHA, nano-SiO_2 and tap water were used. The RHA used in this experiment contained 92.1% SiO_2 with an average particle size of 15.83 μm. The chemical compositions of RHA and cement were determined using an x-ray microprobe analyzer (Table 5.1).

TABLE 5.1 Chemical composition of cement and rice husk ash.

Chemical	Chemical composition (%)	
	OPC	RHA
SiO_2	21	92.1
Al_2O_3	4.6	0.41
Fe_2O_3	3.2	0.21
CaO	64.5	0.41
MgO	2	0.45
SO_3	2.9	-
LOI	1.5	-

In order to achieve the desired fluidity and better dispersion of nanoparticles, the polycarboxylate ether based superplastisiser was incorporated into all mixes. The content of superplastisiser was adjusted for each mixture to keep the fluidity of the mortars constant. Natural river sand was used with the fraction of sand, which passed through a 1.18 mm sieve and was retained on a 0.2 mm sieve (conforming to ASTM C778 [53]). The specific gravity of the sand was 2.51 g/cm³. The basic material properties of the nano-SiO₂ are given in Table 5.2.

TABLE 5.2   The properties of nano-SiO₂ used in this experiment.

Item	Diameter (nm)	Specific surface (m²/g)	Density (g/cm³)	Purity (%)
Target	50	50	1.03	99.9

Ten different combinations as listed in Table 5.3 were cast. The compositions A-1 to A-6 were used to investigate the effect of substituting nano-SiO₂ for cement in the mortar and to determine the optimum content of nano-SiO₂ in cement mortar. The compositions A-7 to A-10 were made to study the effect of nano-SiO₂ particles on the properties of cement mortar containing RHA. The amount of RHA replacement in the mortar was 20% by weight of cement which is an acceptable range and is the one most often used. The water/binder ratio for all mixtures was 0.5 where the binder weight is the total weight of cement, RHA and nano-SiO₂. The cement/sand ratio was 1:2.75 for all the mixtures. Cubes (50 × 50 × 50 mm) were used for compressive and water absorption tests and beams (50 × 50 × 200 mm) were used for flexural and shrinkage tests. The fresh mortar was placed into the molds and tamped using a hard rubber mallet. After 24 h, the specimens were removed from the molds and cured in water at 23 ± 2 °C until they were tested. The samples were tested using a hydraulic testing machine under load control at 1350 N/S for the compressive test (as per ASTM C109/C109 M [54]) and 44 N/S for the flexural test (as per ASTM C348 [55]). After the mechanical tests, the crushed specimens were selected for scanning electronic microscopy (SEM). The absorption test was carried out on two 50 mm cubes. Saturated surface dry specimens were kept in an oven at 110°C for 72 h. After measuring the initial weight, specimens were immersed in water for 72 h, then the final weight was measured and the final absorption was reported to assess the mortar permeability.

TABLE 5.3   Mix proportion for preparation of the specimens.

Batch No	S^a/B^b	W^c/B	% Content (by weight)		
			°C	RHA	NS
A-1	2.75	0.5	100	-	0
A-2	2.75	0.5	99	-	1
A-3	2.75	0.5	97	-	3
A-4	2.75	0.5	95	-	5
A-5	2.75	0.5	93	-	7
A-6	2.75	0.5	91	-	9
A-7	2.75	0.5	80	20	0
A-8	2.75	0.5	79	20	1
A-9	2.75	0.5	77	20	3
A-10	2.75	0.5	75	20	5

[a]Sand.
[b]Binder (cement + RHA + Nano-SiO_2).
[c]Water.

13.6.2   RESULTS AND DISCUSSION

Figure 5.1 shows the variation in the compressive strength of ordinary and RHA cement mortars at various contents of nano-SiO_2. It can be seen that the compressive strength of cement mortars with nano-SiO_2 are all higher than that of plain cement mortar. Results indicate that the optimal content of nano-SiO_2 for reinforcing a concrete/mortar should be about 7%. It is clear that increasing the amount of the nano-SiO_2 from 7% to 9% doesn't have a considerable effect on the compressive strength. Moreover, larger amounts of nano-SiO_2 actually reduce the strength of composites instead of improving it. It was found that when the content of nanoparticles is large, they are more difficult to disperse uniformly. Therefore, they create a weak zone in the form of voids and consequently the homogeneous hydrated microstructure cannot be formed and a low strength is expected. From the results it is clear that nano-SiO$_2$ is more effective than RHA in reinforcement of cement mortar. Also it can be observed that the nano-SiO$_2$ improved the compressive strength of RHA mortars and incorporating of nano-SiO$_2$ particles by RHA in mortar can lead to further improvements in compressive strength and most likely to improvement of other properties of the cement mortar.

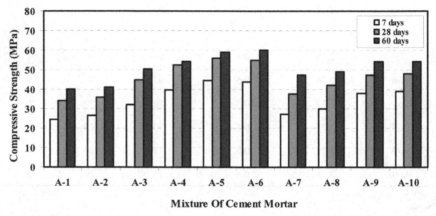

FIGURE 5.1   Compressive strength of the ordinary and RHA cement mortars at different content of nano-SiO$_2$.

The flexural strength of 10 mortar mixtures of different ages is shown in Fig. 5.2. It can be seen that the flexural strengths of the specimens with replacement of cement by nano-SiO$_2$ are all higher than that of plain mortar with the same water to binder ratio. The greatest increase in all ages is observed for batch A4 with 5% nano-SiO$_2$. Using 5% nano-SiO$_2$, the flexural strength at 28 days was 7.5 MPa whereas it decreased to 6.2 MPa with 9% nano-SiO$_2$. It indicates that a high amount of nano-SiO$_2$ (especially in excess of 7%) has a negative effect on flexural strength. It is clear that the nano-SiO$_2$ particles are more effective than RHA in developing flexural strength and incorporating nano-SiO$_2$ in cement mortars containing RHA can further increase the flexural strength. Two fundamental mechanisms can be deduced for strength enhancement by nano-SiO$_2$. The first strengthening mechanism is the filler effect. The micro filling effect of nano-SiO$_2$ is one of the important factors for the development of dense concrete/mortar with a very high strength, because it has been found that small amounts of air significantly decrease the strength of the mortar. It has been reported that the size ratio between filler and the aggregates is one of the main parameters, which strongly affects the strengthening caused by the filling effect. Thanks to the high size ratio between nano-SiO_2 and aggregates, the filling effect of nano-SiO_2 particles is more obvious. Furthermore, the microstructure of the transition zone between the aggregates and the cement paste strongly influences the strength and durability of the concrete. Absence of nano-SiO_2 particles reduces the wall effect in the transition zone between the paste and the aggregates and strengthens this weaker zone due to the higher bond between those two phases. This mechanism also leads to an improvement in microstructure and properties of the mortars/concretes. The second strengthening mechanism is the pozzolanic activity. Two major products of cement hydration are CSH and CH. CSH which is produced by hydration of C_3S and C_2S plays a vital role in the mechanical characteristics of the cement paste, whereas calcium hydrate which is also formed by hydration of cement does not have any cementing property. It contains about 20–25% of the volume of the hydration products. Calcium hydrates due to their morphology are relatively weak and brittle. Cracks can easily propagate through regions populated by them, especially at the aggregate/cement paste interface. Nano-SiO_2 particles react rapidly with the calcium hydrates formed during hydration of cement and produce CSH with cementitious properties which is beneficial for enhancement of the strength of concrete/mortar.

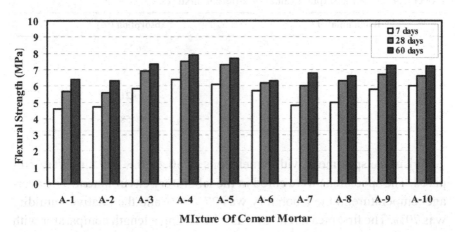

FIGURE 5.2 Flexural strength of the ordinary and RHA cement mortars at different content of nano-SiO_2.

The absorption characteristics indirectly represent the porosity through an understanding of the permeable pore volume and its connectivity. In order to investigate the effect of nano-SiO_2 particles on cement mortar permeability, a water absorption test was carried out on various mixtures A-1 (plain cement mortar), A-5 (cement mortar with 7% nano-SiO_2), A-7 (cement mortar with 20% RHA replacement) A9 (cement mortar with 20% RHA replacement and 3% nano-SiO_2). The final absorption of these mixtures are shown in Table 5.4. It can be seen that mixture A-5 (cement mortar with 7% nano-SiO_2) showed the lowest absorption of all the mixtures which shows that nano-SiO_2 is more effective in reducing the permeability than RHA. Integrating nano-SiO_2 into RHA mortar reduced the water absorption from 5.42% to 4.45%. Results showed that the presence of nano-SiO_2 particles in cement mortar could decrease the water absorption and the likely permeability of cement mortar. This impermeability increase can be attributed to two concomitant phenomena:

1. Nano-SiO_2 particles generate a large number of nucleation sites for hydration products and induce a more homogenous distribution of CSH and thus, there are less pores, and

2. Nano-SiO_2 particles block the passages connecting capillary pores and water channels in the cement paste.

TABLE 5.4 Water absorption values of different mixtures.

Batch number	Absorption (%)
A-1	6.12
A-5	4.23
A-7	5.421
A-9	4.458

Prismatic specimens with dimensions of 50 × 50 × 200 mm were prepared. The specimens were cured in the laboratory environment. The average temperature in the laboratory was 27 ± 3°C and the relative humidity was 70%. The first measurement was taken using a length comparator with a precision of 2 μm after 24 h of mixing, while the rest of measurements were taken after 3, 7, 14, 21, 28, 35, 42 days. The shrinkage behavior of the mortars containing nano-SiO_2 is shown in Fig. 5.3.

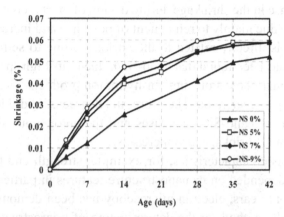

FIGURE 5.3   Shrinkage of mortars containing nano-SiO$_2$ versus time.

From the results, it can be seen that the shrinkage during drying of mortars with nano-SiO$_2$ is apparently higher than that of the control mortar and increases with increasing nano-SiO$_2$ content. Fig. 5.4 shows the influence of nano-SiO$_2$ on the shrinkage behavior of the RHA mortar. Results showed that RHA mortar experienced higher shrinkage than that of ordinary cement mortar. An increase was observed in RHA mortar containing nano-SiO$_2$ compared to the RHA mortar.

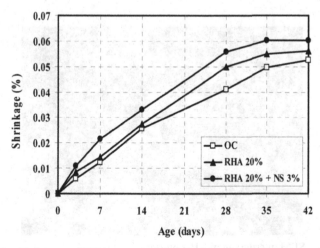

FIGURE 5.4   Shrinkage of RHA mortar with and without nano-SiO$_2$ versus time.

The increase in the shrinkage during drying of mortar containing nano-SiO_2 might be due mainly to refinement of pore size and increase of meso-pores, which is directly related to the shrinkage due to self-desiccation. Moreover, it has been found that nano-SiO_2 due to its high specific surface serves additional nucleation sites for hydration products whereby chemical reactions are accelerated. Therefore, the degree of hydration increases as the amount of nano-SiO_2 increases and the autogenous shrinkage related to chemical shrinkage also increases.

Cement paste characteristics, for example, strength and permeability significantly depended on its nanostructure features in particular nanoporosity. In recent years, electron microscopy has been demonstrated to be a very valuable method for the determination of microstructure. Numerous studies on the influence of nano-SiO_2 on the microstructure of plain cement mortar have been carried out. The results showed that nano-SiO_2 particles formed a very dense and compact texture in the hydrate products and decreased the size of big crystals such as CH. In this chapter in order to study the microstructure of RHA mortar, with and without nano-SiO_2, a SEM was used. The microstructure of the RHA mortar with 3% replacement of nano-SiO_2 and without nano-SiO_2 at a curing age of seven days are presented in Fig. 5.5 and Fig. 5.6, respectively. Results showed that the nano-SiO_2 particles improved the dense and compact microstructure of RHA and generated a more homogenous distribution of hydrated products.

FIGURE 5.5   SEM micrograph of RHA mortar.

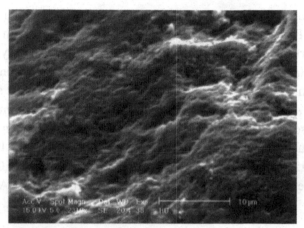

FIGURE 5.6   SEM micrograph of RHA mortar with 3% nano-SiO$_2$.

13.6.3   CONCLUDING REMARKS

The construction industry uses a large amount of concrete. Concrete is used in infrastructure and in buildings. It is composed of granular materials of different sizes and the size range of the composed solid mix covers is large. The overall grading of the mix, containing particles from 300 nm to 32 mm determines the mix properties of the concrete. The properties in fresh state (flow properties and workability) are, for example, governed by the particle size distribution, but also the properties of the concrete in its hardened state, such as strength and durability, are affected by the mix grading and resulting particle packing. One way to further improve the packing is to increase the solid size range, e.g., by including particles with sizes below 300 nm. Possible materials, which are currently available are limestone and silica fines such as silica flavor, silica fume and nano-SiO$_2$. However, these products are synthesized in a rather complex way, resulting in high purity and complex processes that make them nonfeasible for the construction industry.

A noticeable increase was observed in compressive and flexural strength of ordinary cement mortars upon adding nano-SiO$_2$. Compressive and flexural strength of RHA cement mortars improved with the incorporation of nano-SiO_2. Integrating nano-SiO_2 with cement mortar containing RHA improved the microstructure of products in dense and compact forms. Nano-SiO_2 had a significant impact on the drying shrinkage of mortars. The mortar samples with nano-SiO_2 experienced higher drying shrinkage. This effect was more prominent for larger amounts of nano-SiO_2. According to the results there was a significant improvement in water absorption of mortars with integrating nano-SiO_2. Nano-SiO_2 particles decreased the water absorption of the cement composite by pore filling and pozzolanic effects. Also it was observed that nano-SiO_2 particles were more effective in the reduction of permeability than RHA.

KEYWORDS

- **chemicals complexity**
- **control**
- **limitations**
- **nanoparticles**
- **nanopolymers**
- **new approaches**

REFERENCES

1. G.I. Taylor in the *Proceedings of the Royal Society A*, 1969, **313**, 1515, 453.
2. J. Doshi and D.H. Reneker, *Journal of Electrostatics*, 1995, **35**, 2–3, 151.
3. H. Fong and D.H. Reneker in *Structure Formation in Polymeric Fibres*, Hanser Gardener Publications, Cincinnati, OH, USA, 2001, p.225.
4. D. Li and Y. Xia, *Advanced Materials*, 2004, **16**, 14, 1151.
5. R. Derch, A. Greiner and J.H. Wendorff in *Dekker Encyclopedia of Nanoscience and Nanotechnology*, Eds., J.A. Schwarz, C.I. Contescu and K. Putyera, CRC Press, New York, NY, USA, 2004, p.182.
6. A.K. Haghi and M. Akbari, *Physica Status Solidi A*, 2007, **204**, 6, 1830.
7. P.W. Gibson, H.L. Schreuder-Gibson and D. Rivin, *AIChE Journal*, 1999, **45**, 1, 190.
8. Z-M. Huang, Y-Z. Zhang, M. Kotaki and S. Ramakrishna, *Composites Science and Technology*, 2003, **63**, 15, 2223.
9. M. Li, M.J. Mondrinos, M.R. Gandhi, F.K. Ko, A.S. Weiss and P.I. Lelkes, *Biomaterials*, 2005, **26**, 30, 5999.

10. E.D. Boland, B.D. Coleman, C.P. Barnes, D.G. Simpson, G.E. Wnek and G.L. Bowlin, *Acta Biomaterialia*, 2005, **1**, 1, 115.
11. J. Lannutti, D. Reneker, T. Ma, D. Tomasko and D. Farson, *Materials Science and Engineering C*, 2007, **27**, 3, 504.
12. J. Zeng, L. Yang, Q. Liang, X. Zhang, H. Guan, C. Xu, X. Chen and X. Jing, *Journal of Controlled Release*, 2005, **105**, 1–2, 43.
13. E-R. Kenawy, G.L. Bowlin, K. Mansfield, J. Layman, D.G. Simpson, E.H. Sanders and G.E. Wnek, *Journal of Controlled Release*, 2002, **81**, 1–2, 57.
14. M-S. Khil, D-I. Cha, H-Y. Kim, I-S. Kim and N. Bhattarai, *Journal of Biomedical Materials Research, Part B: Applied Biomaterials*, 2003, **67**, 2, 675.
15. B-M. Min, G. Lee, S.H. Kim, Y.S. Nam, T.S. Lee and W.H. Park, *Biomaterials*, 2004, **25**, 7–8, 1289.
16. X-H. Qin and S-Y. Wang, *Journal of Applied Polymer Science*, 2006, **102**, 2, 1285.
17. J-S. Kim and D.H. Reneker, *Polymer Engineering and Science*, 1999, **39**, 5, 849.
18. S.W. Lee, S.W. Choi, S.M. Jo, B.D. Chin, D.Y. Kim and K.Y. Lee, *Journal of Power Sources*, 2006, **163**, 1, 41.
19. C. Kim, *Journal of Power Sources*, 2005, **142**, 1–2, 382.
20. N.J. Pinto, A.T. Johnson, A.G. MacDiarmid, C.H. Mueller, N. Theofylaktos, D.C. Robinson and F.A. Miranda, *Applied Physics Letters*, 2003, **83**, 20, 4244.
21. D. Aussawasathien, J-H. Dong and L. Dai, *Synthetic Metals*, 2005, **154**, 1–3, 37.
22. S-Y. Jang, V. Seshadri, M-S. Khil, A. Kumar, M. Marquez, P.T. Mather and G.A. Sotzing, *Advanced Materials*, 2005, **17**, 18, 2177.
23. S-H. Tan, R. Inai, M. Kotaki and R. Ramakrishna, *Polymer*, 2005, **46**, 16, 6128.
24. A. Ziabicki in *Fundamentals of Fibre Formation: Science of Fibre Spinning and Drawing*, John Wiley & Sons, New York, NY, USA, 1976, p.76.
25. A. Podgóski, A. Bałazy and L. Gradoń, *Chemical Engineering Science*, 2006, **61**, 20, 6804.
26. B. Ding, M. Yamazaki and S. Shiratori, *Sensors and Actuators B*, 2005, **106**, 1, 477.
27. J.R. Kim, S.W. Choi, S.M. Jo, W.S. Lee and B.C. Kim, *Electrochimica Acta*, 2004, **50**, 1, 69.
28. L. Moroni, R. Licht, J. de Boer, J.R. de Wijn and C.A. van Blitterswijk, *Biomaterials*, 2006, **27**, 28, 4911.
29. T. Wang and S. Kumar, *Journal of Applied Polymer Science*, 2006, **102**, 2, 1023.
30. S. Sukigara, M. Gandhi, J. Ayutsede, M. Micklus and F. Ko, *Polymer*, 2004, **45**, 11, 3701.
31. S.Y. Gu, J. Ren and G.J. Vancso, *European Polymer Journal*, 2005, **41**, 11, 2559.
32. S-Y. Gu and J. Ren, *Macromolecular Materials and Engineering*, 2005, **290**, 11, 1097.
33. O.S. Yördem, M. Papila and Y.Z. Menceloğlu, *Materials and Design*, 2008, **29**, 1, 34.
34. I. Sakurada, *Polyvinyl Alcohol Fibres*, Marcel Dekker, New York, NY, USA, 1985, p.167.
35. F.L. Marten in *Encyclopedia of Polymer Science and Technology*, Ed., H.F. Mark, 3rd Edition, Volume 8, John Wiley & Sons, New York, NY, USA, 2004, p.154.
36. Y.D. Kwon, S. Kavesh and D.C. Prevorsek, inventors; Allied Corporation, assignee; US 4,440,711, 1984.
37. S. Kavesh and D.C. Prevorsek, inventors; Allied Corporation, assignee; US 4,551,296, 1985.

38. H. Tanaka, M. Suzuki and F. Uedo, inventors; Toray Industries, Inc., assignee; US 4,603,083, 1986.
39. G. Paradossi, F. Cavalieri, E. Chiessi, C. Spagnoli and M.K. Cowman, *Journal of Materials Science: Materials in Medicine*, 2003, **14**, 8, 687.
40. Z-Q. Gu, J-M. Xiao and X-H. Zhang, *Bio-Medical Materials and Engineering*, 1998, **8**, 2, 75.
41. M. Oka, K. Ushio, P. Kumar, K. Ikeuchi, S.H. Hyon, T. Nakamura and H. Fujita, *Journal of Engineering in Medicine*, 2000, **214**, 1, 59.
42. K. Burczak, E. Gamian and A. Kochman, *Biomaterials*, 1996, **17**, 24, 2351.
43. J.K. Li, N. Wang and X.S. Wu, *Journal of Controlled Release*, 1998, **56**, 1–3, 117.
44. A.S. Hoffman, *Advanced Drug Delivery Reviews*, 2002, **43**, 3.
45. J. Zeng, A. Aigner, F. Czubayko, T. Kissel, J.H. Wendorff and A. Greiner, *Biomacromolecules*, 2005, **6**, 3, 1484.
46. K.H. Hong, *Polymer Engineering and Science*, 2007, **47**, 1, 43.
47. L.H. Sperling, *Introduction to Physical Polymer Science*, 4th Edition, John Wiley & Sons, Hoboken, NJ, USA, 2006, p.76.
48. J.C.J.F. Tacx, H.M. Schoffeleers, A.G.M. Brands and L. Teuwen, *Polymer*, 2000, **41**, 3, 947.
49. F.K. Ko in *Nanomaterials Handbook*, Ed., Y. Gogotsi, CRC Press, Boca Raton, FL, USA, 2006, Chapter 19.
50. A. Koski, K. Yim and S. Shivkumar, *Materials Letters*, 2004, **58**, 3–4, 493.
51. D.C. Montgomery, *Design and Analysis of Experiments*, 5th Edition, John Wiley & Sons, New York, NY, USA, 1997, p.95.
52. A. Dean and D. Voss, *Design and Analysis of Experiments*, Springer, New York, NY, USA, 1999, p.129.
53. G.E.P. Box and N.R. Drape, *Response Surfaces, Mixtures, and Ridge Analyzes*, 2nd Edition, John Wiley & Sons, Hoboken, NJ, USA, 2007, p.94.
54. K.M. Carley, N.Y. Kamneva and J. Reminga, *Response Surface Methodology*, CASOS Technical Report, CMU-ISRI-04–136, 2004.
55. S. Weisberg, *Applied Linear Regression*, 3rd Edition, Wiley, Hoboken, NJ, USA, 2005, p.69.
56. C. Zhang, X. Yuan, L. Wu, Y. Han and J. Sheng, *European Polymer Journal*, 2005, **41**, 3, 423.
57. Q. Li, Z.D. Jia, Y. Yang, L.M. Wang and Z.C. Guan in the *Proceedings of the 7th IEEE International Conference on Solid Dielectrics*, Winchester, UK, 2007, p.215.
58. C. Mit-uppatham, M. Nithitanakul and P. Supaphol, *Macromolecular Chemistry and Physics*, 2004, **205**, 17, 2327.
59. T. Jarusuwannapoom, W. Hongrojjanawiwat, S. Jitjaicham, L. Wannatong, M. Nithitanakul, C. Pattamaprom, P. Koombhongse, R. Rangkupan and P. Supaphol, *European Polymer Journal*, 2005, **41**, 3, 409.
60. S.C. Baker, N. Atkin, P.A. Gunning, N. Granville, K. Wilson, D. Wilson and J. Southgate, *Biomaterials*, 2006, **27**, 16, 3136.
61. S. Sukigara, M. Gandhi, J. Ayutsede, M. Micklus and F. Ko, *Polymer*, 2003, **44**, 19, 5721.

NOMENCLATURES

A	variant, Kuhn segment length
A_p	dependence
a	lower linear scale of fractal behavior
AFM	atomic force microscope
ASTM	American Society for Testing and Materials
b_B	Burgers vector value
BSR	butadiene-styrene rubber
C_∞	characteristic ratio
c	nanoparticles concentration
c_0	"seeds" number [*equals to nanoparticles clusters (aggre gates) number*]
CH	calcium hydroxide crystals
CSH	hydrated calcium silicate
d	dimension of Euclidean space
d_f	fractal dimension
d_{surf}	fractal dimension of nanofiller surface
D	diffusivity
D_n	fractal dimension
D_p	particles diameter
E	Young modulus
E_m	elasticity moduli of matrix polymer
E_n	elasticity moduli of nanocomposite
E_0	initial modulus
EP/MMT	epoxy polymer/Na^+-montmorillonite
G	Grüneisen parameter, shear modulus in solid-phase
G_0	shear modulus of matrix polymer
HDPE	high density polyethylene
k	Boltzmann constant
k_n	proportionality coefficient
K	bulk elasticity modulus
K_T	isothermal modulus of dilatation
K_m	bulk elasticity modulus of polymeric matrix
K_n	bulk elasticity modulus of nonofiller
l	a linear scale for fractals interpenetration distance

l_{if}	interfacial layer thickness
l_0	the main chain skeletal bond length
l_{st}	statistical segment length
l_{if}^T	theoretical value of interfacial layer thickness
L_n	mean distance between nanofiller particles aggregates
MFI	melt flow index
MPa	megapascal
N	nanoparticles number per one aggregate
N_i	quadrates number
NC's	nylon-6/clay nanocomposites
NR	natural rubbers
OSY	oxinitride silicium-yttrium
P	load
P-Cl	particle-cluster
PLSN	polymer/layered silicate nanocomposites
PP	polypropylene
PP-g-MA	polypropylene-graft-maleic anhydride
PSD	particle size distribution
r_n	nanoparticles radius
R_{ag}	average discrepancy
R_{max}	nanoparticles cluster
R_p	nanofiller particle radius
RHA	rice husk ash
S	macromolecule cross-sectional area
S_i	quadrate area
S_n	cross-sectional area of nanoparticles
S_u	nanoshungite particles specific surface, specific surface of aerosil particles
SEM	scanning electron microscopy
Sf	silica flavor
SF	silica fume
SPIP	Scanning Probe Image Processor
SPM	Scanning Probe Microscope
STM	Scanning Tunneling Microscope
t	duration, percolation index
T	temperature

T_g	glass transition temperature
TC	technical carbon
TGA	thermo-gravimetric analysis
W_n	nanofiller mass content
WAXD	wide-angle X-ray scattering
XRD	X-ray diffraction
α_i	quadrate side size
α_m	thermal expansion coefficient of polymeric matrix
$\alpha_{n\,m}$	thermal expansion coefficient of nanofiller
α_n^m	thermal expansion coefficient of mixture
ρ	particles size
ρ_n	nanofiller particles aggregate density
φ_{if}	relative volume fractions of interfacial regions
φ_n	relative volume fractions of nanofiller, nanofiller volume contents
c_{AB}	Flory-Huggins interaction parameter
h	a packing coefficient, medium viscosity
η_0	initial viscosity, constant
φ_{den}	relative fraction of nanocomposites densely packed regions
φ_n	nanofiller volume contents, nanofiller volume fraction
a	numerical coefficient
x	correlation length of diffusion
ζ	walker's diffusion constant
σ	normal stress
τ	shear stress
l	distance between nonaggregated nanofiller particles, wave length
g_L	Grüneisen parameter
n	Poisson ratio
v_{TC}	nanofiller (technical carbon) Poisson's ratio
v_m	polymer matrix Poisson's ratio
Δ_f	Hausdorff dimension

INDEX

Printed in the United States
by Baker & Taylor Publisher Services